Clinical Examination of the Musculoskeletal System: Assessing Rheumatic Conditions

Clinical Examination of the
Musculoskeletal System:
Assessing Rheumatic Conditions

The illustration on the front cover is by the famous Hungarian artist, Jenó Barcsay (1900–1988), who was Professor of the Hungarian Academy of Fine Arts. He lived and worked in the artists' colony in Szentendre, known as the attractive town of the museums. Szentendre is 26 kilometers from Budapest and has been described as the Montmartre of the Danube. In the introduction of the 13th edition of his book, *Anatomy for the Artist* (published in 1982 with an English translation by Octopus Books Limited in London, England), he makes the comment:

> *"the artist should therefore approach anatomy from an artist's point of view, not a doctor's. He should make himself acquainted primarily with the bones and muscles whose forms and actions are visible at just below the surface. The study of the heart, liver and lungs will improve his general knowledge but not his ability to draw. Therefore* Anatomy for the Artist *does not deal with the internal organs."*

In short, it deals with the musculoskeletal system and has become a key text for an entire generation of artists.

Clinical Examination of the Musculoskeletal System: Assessing Rheumatic Conditions

W. Watson Buchanan, M.D., F.R.C.P. (Glasg. and Edin), F.A.C.P., F.A.C.R. (Hon)

Emeritus Professor of Medicine
McMaster University, Faculty of Health Sciences
Consultant Rheumatologist
Sir William Osler Institute of Health
Hamilton, Ontario, Canada

Karel de Ceulaer, M.D.

Honorary Lecturer and Consultant Rheumatologist
Department of Medicine
University of West Indies
Jamaica, West Indies

Géza P. Bálint, M.D., D.Sc.

Professor of Physiotherapy
Haynal Imre University, Faculty of Health Sciences
Director, National Institute of Rheumatology and Physiotherapy
Budapest, Hungary

Williams & Wilkins
A WAVERLY COMPANY
BALTIMORE • PHILADELPHIA • LONDON • PARIS • BANGKOK
BUENOS AIRES • HONG KONG • MUNICH • SYDNEY • TOKYO • WROCLAW

Editor: Jonathan W. Pine, Jr.
Managing Editor: Leah Ann Kiehne Hayes
Production Coordinator: Marette Magargle-Smith
Copy Editor: Lisa Keisel
Designer: Silverchair Science & Communication Inc.
Illustration Planner: Lorraine Wrzosek
Cover Designer: Silverchair Science & Communication Inc.
Typesetter: Peirce Graphics Services, Inc.
Printer: Vicks Lithograph & Printing Corporation
Binder: Vicks Lithograph & Printing Corporation

Accurate indications, adverse reactions, and dosage schedules for drugs are provided in this book, but it is possible that they may change. The reader is urged to
review the package information data of the manufacturers of the medications mentioned.

Printed in the United States of America

First Edition.

Library of Congress Cataloging in Publication Data

Buchanan, W. Watson.
    Clinical examination of the musculoskeletal system / W. Watson
Buchanan, Géza P. Bálint, Karel de Ceulaer.
        p.   cm.
    Includes index.
    ISBN 0-683-01127-8
    1. Musculoskeletal system—Examination.   I. Bálint, Géza.
II. Ceulaer, Karel de.   III. Title.
    [DNLM: 1. Rheumatic Diseases—diagnosis.   2. Musculoskeletal
System—patology.   WE 544   B918c   1996]
RC925.7.B83   1996
616.7'075—dc20
DNLM/DLC
for Library of Congress   96-7787
                                                          CIP

*The publishers have made every effort to trace the copyright holders for borrowed ma-
terial. If they have inadvertently overlooked any, they will be pleased to make the necessary arrangements at the first opportunity.*

To purchase additional copies of this book, call our customer service department at
(800) 638-0672 or fax orders to **(800) 447-8438.** For other book services, including
chapter reprints and large quantity sales, ask for the Special Sales department.

Canadian customers should call **(800) 268-4178,** or fax **(905) 470-6780.** For all
other calls originating outside of the United States, please call **(410) 528-4223** or
fax us at **(410)528-8550.**

Visit *Williams & Wilkins* on the Internet: **http://www.wwilkins.com** or contact our
customer service department at **custserv@wwilkins.com.** Williams & Wilkins customer service representatives are available from 8:30 am to 6:00 pm EST, Monday
through Friday, for telephone access.

97 98 99
1 2 3 4 5 6 7 8 9 10

"If thou examines a man having a sprain in a vertebra of his spinal column, thou shouldst say to him: 'extend now thy two legs, and contract them both again.' When he extends them both, he contracts them immediately, because of the pain he causes in the vertebra of his spinal column in which he suffers."

Possible description of Lasègues sign in the Edwin Smith surgical papyrus of ancient Egypt.

Foreword

I am not a specialist in diseases of the locomotor system and, therefore, cannot attest to the contents of this book. However, not being a specialist allows me to appreciate the approach of the book from the point of view of a student. I am particularly impressed by three features. First, the overall organization. Although they recognize the importance of the "seven pillars," the authors do not recommend hidebound rituals in their deployment. For example, when appropriate examination is given precedence over history. Secondly, the approach is problem-oriented rather than disease-oriented. Thirdly, the illustrations are excellent and relate well to the text.

I have enjoyed reading this book and am confident it will appeal to students, residents, family practitioners, general internists, and many other specialists including, perhaps, those dealing with the locomotor system.

E.J.M. Campbell
Founding Chairman, Department of Medicine
Faculty of Health Sciences
McMaster University
Hamilton, Ontario, Canada

Preface

First Witch:	When shall we three meet again?
	In thunder, lightening, or in rain?
Second Witch:	When the hurlyburly's done, when the battle's
	lost and won.
Third Witch:	That will be ere the set of sun.
First Witch:	Where the place?
Second Witch:	Upon the heath.
(Act 1. Scene 1. "Macbeth" by William Shakespeare)	

Unlike the three witches in the play "Macbeth," we did not meet on a lonely, wind-swept moor, but at the Center for Rheumatic Diseases in Glasgow, Scotland in the 1970s. At that time, we felt there was a need for a concise and compact account of clinical examination of the locomotor system. Although we went our separate ways, we continued to maintain the need for such a book.

This book summarizes what we consider important for the physician in clinical examination of a patient with arthritis or an allied condition. We included laboratory tests and radiologic investigations because we believe they form an integral part of clinical assessment. In the appendix, we also included some everyday facts and commonly used references. We hope this book will be useful to medical undergraduates and postgraduates in internal medicine, rheumatology, orthopaedics, and family practice training programs.

International cooperation in medical research has long been advocated in rheumatology. Coming from different cultural backgrounds, as we do, might be considered a handicap in writing a book such as this one on clinical examination. However, the reverse has been true.

W. Watson Buchanan, M.D., F.R.C.P., F.A.C.P., F.A.C.R.
Géza P. Bálint, M.D., D.Sc.
Karel de Ceulaer, M.D.

Introduction

An unhurried, detailed history and careful physical examination are perhaps the most important diagnostic aids that the family and hospital practitioner have in establishing a diagnosis of rheumatic disease. In approximately 40% of patients, only a symptomatic diagnosis, such as myalgia or lumbago, can be made. In the remaining 60% of patients, a more definitive pathologic diagnosis should be possible, such as systemic lupus erythematosus or gout. A thorough history and physical examination establish the diagnosis and enable the doctor to get to know the patient, which in turn facilitates management. A number of internationally recognized criteria have been established for certain diseases [e.g., rheumatic fever (Jones criteria) and rheumatoid arthritis (American Rheumatism Association)] Although these are helpful for research purposes, they do not aid the family physician in establishing a diagnosis and prescribing therapy for an individual patient, the primary concern of the family physician.

T.E. Lawrence (1888–1935), otherwise known as Lawrence of Arabia, popularized the "seven pillars of wisdom" from the book of Proverbs, chapter IX, verse 1. In medicine, the "seven pillars of wisdom" for establishing a diagnosis are:

1. History
2. Physical examination
3. Radiographic examination
4. Laboratory tests
5. Histology
6. Consultation with other specialists
7. The outcome of the illness

Consultant physicians have been known to diagnose a patient as having rheumatoid arthritis on the basis of an inflammatory erosive polyarthritis with a negative test for rheumatoid factor and changed the diagnosis when the patient developed psoriasis. A diagnosis of an obscure polyarthritis may only be established after consultation with a gastroenterologist in the case of regional ileitis or with an ophthalmologist in the case of asymptomatic iridocyclitis associated with anklyosing spondylitis.

A detailed history should be obtained from a patient presenting with problems of the locomotor system. The experienced clinician frequently combines history taking and physical examination, which most textbooks fail to appreciate. For example, a patient with a painful toe takes off his shoe and sock to show the doctor his toe. History and physical examination then follow simultaneously. If gout is suspected, the doctor frequently performs an aspiration to demonstrate crystals. Having established the diagnosis, the doctor excludes secondary causes of gout. Medical students are taught in an artificial environment and writers of medical texts continue to propagate a dull, stereotyped, and time-consuming exercise in history taking and physical examination. Therefore, we placed physical examination before history taking.

Contents

Physical Examination in General

1

INSPECTION

It is important in all clinical specialties to first inspect the patient as a whole. Note should be taken of posture, gait and general ability to move, and the presence of gross joint deformities and mucocutaneous lesions. The following features should be noted during the physical examination.

Posture

When examining posture, it is important to notice any abnormal positioning of the head and neck. If the neck is bent forward and stiff, ankylosing spondylitis may be present. The neck bent forward without stiffness may indicate myopathy or myasthenia. Torticollis may result when the patient tries to avoid pain caused by a variety of conditions, such as cervical injury or disease, muscle spasm, congenital torticollis, muscle weakness, nerve injury, or myopathies. The need to support the head with the hands may indicate high-level fractures, cervical spine disease, or myopathies.

Level of the Shoulders

During the physical examination it is also important to observe the level of the patient's shoulders. The patient may hold one shoulder higher or lower than the other. An elevated shoulder may be due to muscle spasms caused by painful cervical disease or a nerve lesion. An elevated shoulder may also be the result of a displaced scapula (e.g., Sprengel deformity). A drooping shoulder may be due to muscle weakness (e.g., unilateral myopathy, 11th nerve palsy) or bone disease (e.g., upper thoracic scoliosis).

Spinal and Thoracic Deformities

If the examination reveals the presence of scoliosis, the possibility of reducing the scoliosis should be evaluated. Scoliosis is functional if it can be corrected by sitting, due to shortening of a leg, or lying, due to muscle weakness or avoiding pain. Scoliosis that cannot be reduced is most likely due to organic lesions of the spine (e.g., injury, congenital scoliosis).

A kyphosis may also be found during the physical examination. The kyphosis may be functional, due to avoidance of pain in the spine or abdomen (e.g., pancreatic pain). It may also be compensatory, due to contracture of hips and knees or abdominal muscle contractures after surgery. Organic kyphosis may exist with angulation (e.g., Pott's disease, vertebral body collapse) or without angulation (e.g., ankylosing spondylitis).

Chest wall deformities may be helpful in making an overall diagnosis (e.g., funnel chest in Marfan's syndrome and "rosary" in childhood rickets).

Leg Length Differences

Apparent shortening of a limb may be due to a pelvic tilt. Absolute shortening of a limb may be due to severe joint destruction in the hip or knee, trauma to bones, or deformities of the feet.

Gross Bone and Joint Deformities

Contractures of joints are generally obvious. Deformities of bone may result in a visually obvious diagnosis ("Blick" diagnosis) of Paget's disease due to the enlarged head, kyphosis, and simian stance. In addition, monostotic Paget's disease of the tibia may be immediately recognized.

Although the saber tibia of syphilis no longer exists, deformities due to rickets may still be seen, especially in the older population of Glasgow. Currently, varus deformity developing in the leg of an elderly person is often due to osteoarthritis or calcium pyrophosphate deposition disease.

Gait and Mobility

During the physical examination it is important to ascertain whether the patient can stand and move normally. For example, the patient may limp, require a walking stick, or have a waddling gait. If abnormalities are present, it is necessary to determine which joints are preventing normal movement. The patient's ability to stand on tiptoe or the back of the heels may indicate which joints are responsible.

It is also important to observe the patient rising from a chair. Difficulty associated with this movement may be due to muscular weakness, hip or knee joint arthritis, or spinal disease. When hip and knee joint arthritis are present, the first movement after rest is often the most severe. However, this has no diagnostic specificity because it occurs in osteoarthritis and rheumatoid arthritis.

The patient should also be observed bending forward and note should be taken of any back stiffness. Patients with spinal disease pick up objects from the floor by bending their knees rather than their spines. This is a useful test for spinal disease in children. The test is not as successful in adults because they generally lack the flexibility to perform the required movements (Fig. 1–1).

Movements of the head, shoulder, arms, and hands should also be noted when examining for gait and mobility.

Mucocutaneous Lesions

Diverse manifestations may be observed in the skin and subcutaneous tissues. For example, parotid or submandibular gland enlargement may be seen in Sjögren's syndrome, temporal artery involvement may be seen in cranial arteritis, and alopecia with typical brush border effect may be seen in systemic lupus erythematosus. Also, the skin lesions of psoriasis and scleroderma may be apparent. Recession of the chin (micrognathia or birdlike facies) is present when the temporomandibular joint is involved in juvenile rheumatoid arthritis.

PALPATION

The finger tips are to the rheumatologist what the tendon hammer is to the neurologist and the stethoscope is to the cardiologist. Palpation can probably give more information than any other clinical method in examination of the locomotor system. Palpation can determine whether pain is originating from the skin, subcutaneous tissue, muscle, tendon, joint, bone, or nerve. The patient often helps to locate the site of pain by verbally indicating when the doctor palpates the tender region. The rheumatologist can also use the patient's response to pain arising from palpation to form an impression of the patient's personality.

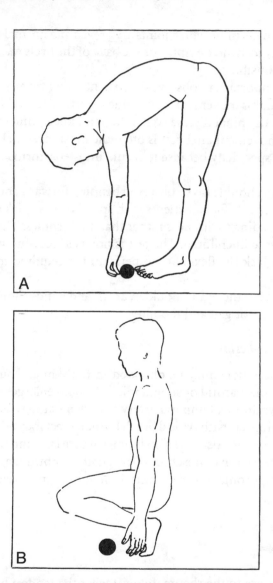

Figure 1–1. Child with painful spine picks up object from floor by bending knees and keeping the spine fixed and straight (B) in contrast to healthy child (A).

Palpation of the joints is essential in examination of the locomotor system. Joint tenderness should be elicited by firm digital pressure along the margins of the joint. Joint tenderness may be the first and only sign of joint inflammation. Joint tenderness can be recorded as zero (patient reports no pain), +1 (patient reports pain), +2 (patient reports pain and grimaces), and

+3 (pain is so severe that the patient withdraws the limb). Using this grading system, an articular index of joint tenderness can be derived and has been useful in monitoring the effects of antirheumatic drugs in clinical therapeutic trials (1).

Palpation may also reveal tenderness in precisely predictable areas (Fig. 1–2) in patients with fibrositis syndrome. The patients often withdraw when these areas are palpated. This type of reaction is called the "jump sign."

Palpation of joints may also be useful in determining synovial tissue hypertrophy and periarticular joint swelling. The synovium has a "boggy" feel and can be distinguished with palpation. Elicitation of an effusion in the knee joint can be achieved by demonstration of a patellar tap or by milking the fluid from one side of the joint to the other (Fig. 1–3).

Osteophytes may be palpated in joints with osteoarthritis. Juxta-articular cysts, such as cysts of the semimembranous bursa (Baker's cyst), behind the knee may also be palpated. If the patient is not in a prone position when examining the popliteal fossa, small cysts may be missed. Popliteal cysts tracking into the calf are more easily palpated when the patient is in the standing position.

Palpation of tendons is important to identify tenosynovitis and the presence of nodules within the tendons. Immediate removal of synovial hypertrophy and nodules may prevent tendon rupture. Fluid in tendon sheaths can

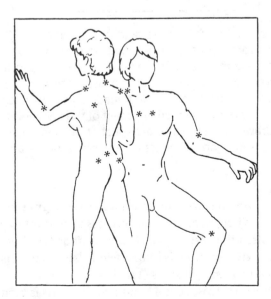

Figure 1–2. Location of 13 typical "trigger" sites of deep tenderness in fibrositis (after Dr. Hugh A. Smythe of University of Toronto).

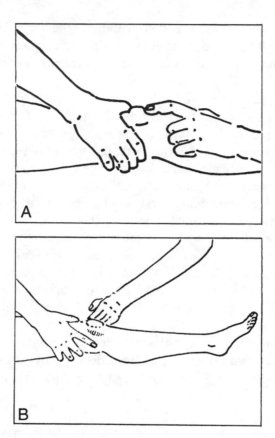

Figure 1–3.   Examination for excess fluid in knee joint. (A) Patellar tap. Synovial fluid from suprapatellar pouch is first squeezed into the knee joint as shown. The patella is then tapped against the femoral condyles. (B) The knee is placed in extension and fluid is again milked into the knee joint from the suprapatellar pouch. The fluid is then forced to the later aspect of the joint by drawing the hand down the medial aspect of the knee joint. When the back of the hand is drawn down the lateral aspect of the joint, fluid is forced to the medial side of the joint and appears as a ripple or bulge as indicated.

also be palpated. Melon seeds can occasionally be diagnosed in a compound palmar ganglion. When palpating tendons, it is important to put them on the stretch. This is often the best way to elicit low-grade tendinitis. For example, stretching of the extensor pollicis longus (Finkelstein's test) is useful in diagnosing de Quervain's tenosynovitis (Fig. 1–4).

In rheumatoid arthritis, dorsal hand tenosynovitis is readily apparent. It is important to examine the flexor tendon sheaths in the palm and fingers by

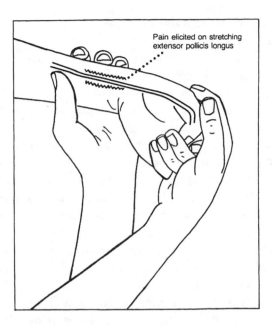

Pain elicited on stretching
extensor pollicis longus

Figure 1–4. Finkelstein's test. Pain is elicited when the extensor pollicis longus is put on the stretch when tenosynovitis is present.

simple palpation with the thumb. A trigger finger can often be diagnosed in this way.

Rupture of tendons is relatively common in rheumatoid arthritis, especially in the extensor tendons of the fingers. Ruptured tendons can be diagnosed based on the history, presence of local tenderness (e.g., in the Achilles tendon), and absence of active movement.

In addition to atrophy, hypertrophy, weakness, spasm, increased tone, and rigidity of skeletal muscle, palpation may reveal generalized or localized tenderness of muscles. It is difficult to differentiate muscle and bone tenderness. Therefore, considerable care must be taken when examining muscle for tenderness. Muscle pain varies in quality and may be described as a dull, boring, or aching sensation. Although fibrositic nodules may be tender when palpated, there is no histological evidence of any specific pathology associated with these nodules.

Generalized bone pain may exist in a number of diverse conditions such as osteomalacia, osteoporosis, hyperparathyroidism, multiple myeloma, and secondary metastases. Generalized bone pain can be elicited by percussion over the spine and pressure on the ribs and iliac crests. Localized bone pain may be elicited by palpation or percussion over the involved bone. Localized pain in the ribs or vertebrae is most often caused by infection or tumor. Pain in juxtaarticular bone may be difficult to differentiate from joint pain. For example, osteomyelitis at the lower end of the femur may simulate an acute arthritis of the knees, especially when an effusion occurs in the joint. The only way to differentiate these two conditions is to determine whether pain is arising from a bone or joint. Palpation of bones may elicit swelling due to tumors, subperiosteal hematomata, or exostosis.

It is not clear why palpation of peripheral nerves elicits pain, since there are no pain receptors in the nerves themselves. However, in peripheral neuropathy, the myelin sheath disappears and the exposed nerve fibers have increased excitability. Palpation of peripheral nerves is important in clinical practice. Hypertrophy of peripheral nerves may be seen in the rare condition of congenital peripheral nerve hypertrophy. A neuroma may also be palpated in a peripheral nerve. A digital nerve neuroma trapped between the two heads of the metatarsal of the foot (Morton's metatarsalgia) may be diagnosed by lateral pressure on the forefoot.

Pressure in a sensory or mixed nerve elicits neurologic pain and paraesthesia in the distribution of the nerve. Putting a nerve on the stretch is useful

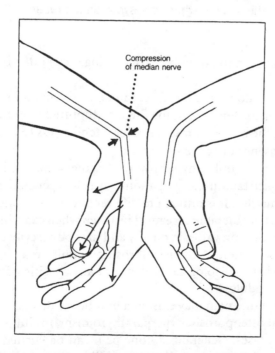

Figure 1–5. Phalen's test to elicit symptoms of carpal tunnel syndrome. The patient's wrist is flexed to its maximum and held for at least one minute.

in eliciting nerve compression. Perhaps the two most common tests involving nerve stretching are straight-leg raising for sciatic pain and Phalen's test for median nerve compression in carpal tunnel syndrome (Fig. 1–5).

Palpation of vessels is as important to the rheumatologist as it is to the cardiologist. For example, examination of the superficial temporal artery may reveal a giant-cell arteritis. Absence of a pulse in the carotids and upper limbs may reveal an aortic arch syndrome (Takayasu's arteritis). A radial pulse, which becomes weak with movement of the head or elevation of the arm, can be characteristic of thoracic outlet syndromes. When polyarteritis nodosa is present, nodules may be felt in peripheral arteries. A glomus tumor in digital vessels may be the cause of a painful finger. The tumor is usually small (the size of a pin head) and red and causes extreme pain with pressure. Ischemic pain from arterial disease at the aortic bifurcation and in the iliac and femoral vessels may cause buttock and leg pain. Absence of femoral, popliteal, anterior, and posterior tibial artery pulses may also be found.

Superficial or deep venous thrombosis may occur in systemic lupus erythematosus, Behçet's syndrome, and visceral neoplasm, all of which may be accompanied by arthritis. A ruptured knee joint or Baker's cyst may closely simulate a deep venous thrombosis in the leg. In this case, diagnosis is based on history rather than physical examination.

FUNCTION

It is important to examine the range of movement of the spine and peripheral joints. Movement of the spine and peripheral joints should be elicited passively and actively. There may be marked limitation in active movement due to muscular weakness when a full range of passive movements can be achieved. The reader is referred to subsequent chapters for a discussion of spine and joint movement in relation to diseases that affect such movement.

REFERENCES

1. Ritchie DM, Boyle JA, McInnes JM, et al. Clinical studies with an articular index for the assessment of joint tenderness in patients with rheumatoid arthritis. Quart J Med 1968;37:393–406.

Patient History | 2

PAIN

Pain is the most important symptom of locomotor disease. It is important to determine the timing, character, and location of pain. These characteristics of pain can be determined by asking a variety of questions. Did the pain begin after an injury occurred? Is this the first attack or a repeat attack of pain? What was the nature and speed of development of the pain? How long has the pain been present and has it been constant or intermittent? Is the pain only felt on movement or is it also present at rest or at rest after exercise? Does the patient have pain at night? Nocturnal pain in hip disease is invariably associated with a joint effusion.

The site of pain is important and may be localized quite accurately by the patient and confirmed by the doctor. It is also important to determine the radiation of pain, particularly in acute lumbar disc protrusion and hip disease. Due to the fact that branches of the obturator and femoral nerves innervate both hip and knee joints, pain from hip joint disease may only be felt in the knee. It is important for the clinician to understand that referred pain from the hip to the knee may be associated with tenderness of the knee. Only careful examination of the hip joint will elicit the true cause of a painful knee. Hip pain may also be referred from the upper lumbar spine (L2 or L3). In L3 nerve root lesions, the pain is often experienced down the anterior aspect of the thigh and becomes more severe with coughing and sneezing. Sacroiliac joint pain, intraabdominal disease (e.g., appendicitis), intrapelvic disease, femoral hernia, and inguinal lymphadenopathy can be manifested as pain in the hip. Referred pain also occurs when there is atlantoaxial dislocation in rheumatoid arthritis, in which the transverse ligament of the axis is put on the stretch. In this case, the patient experiences pain in the occiput and a band-like pain on the forehead. Patients may incorrectly identify pain with an anatomical site. For example, we recently saw a patient with a painful temporomandibular joint due to temporal arteritis.

The character of pain is also important to consider. For example, referred pain is of a diffuse and ill-defined nature, whereas pain from a peripheral nerve or nerve root is often lancinating. The character of pain is difficult to describe irrespective of its origin. Paradoxically, the more intelligent the patient the less clear the description of the pain becomes. This is true for musculoskeletal pain and pain arising from visceral sites. The clinician should try to determine the nature of the pain (e.g., burning, aching, lancinating, throbbing). It is also important to know if there are any aggravating factors. For example, coughing increases pain in lumbar disc protrusion and opposition of the thumb exacerbates pain in a de Quervain's tenosynovitis of the extensor pollicis longus. The clinician should also try to get information about anything that may relieve the pain. For instance, rest or antirheumatic drugs may relieve pain. The position of rest that relieves pain can often be diagnostic. For example, pain associated with acute bursitis of the supraspinatus bursa is often relieved by supporting the elbow with the opposite hand (Fig. 2–1).

It is also useful for the clinician to determine how much disability is due to pain and how much is due to anatomical dysfunction. For example, a painful knee joint may be held in acute flexion and make walking difficult for the patient. Relief of pain may correct the walking problem. On the other hand, the joint may be severely destroyed but not particularly painful (e.g., neuropathic arthropathy). In this case, damage to the joint and not pain is the cause of the patient's disability.

Pain is a subjective experience. The patient's emotional state influences the perceived severity of and reaction to pain. A patient's reaction to pain often helps the clinician to judge the patient's likely response to illness. This is useful in the management of patients with chronic arthritis.

Applying pressure to the midforearm or midjoint of the clavicle can be used to ascertain a patient's pain threshold. It is essential that no pathology exists at these sites, which would render them tender. Many physicians still believe that placebos can be used to determine how much of a patient's pain is genuine. This is incorrect since placebos are effective in diminishing pain from organic disease.

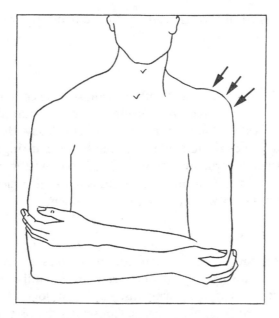

Figure 2–1. A patient with a painful shoulder often obtains relief by supporting the elbow.

JOINTS

Joint Pain

Pain receptor nerve endings are only found in certain joint structures, including the fibrous capsule, fat pads, ligaments, and adventitious sheaths of blood vessels. There are no nerve endings in synovium or cartilage. Therefore, biopsy of these structures can be performed by arthroscopy without an anaesthetic.

It is necessary to distinguish mechanical causes of joint pain from pain due to inflammation. Mechanical causes include increased intraarticular pressure with distension of the capsule, ligamentary subluxation, and trauma to articular fat pads. These types of mechanical joint pain are minimally influenced by nonsteroidal antiinflammatory drugs, including aspirin. On the hand joint, pain arising from synovial inflammation or other causes responds to nonsteroidal antiinflammatory analgesics. This is due to the ability of these drugs to inhibit cyclooxygenase in the tissues.

Those interested in further reading on the neurology of joints are referred to the review by Wyke (1).

Joint Swelling

When taking a history of a patient with musculoskeletal complaints, it is important to determine whether joint swelling has ever occurred. A history of joint swelling makes an inflammatory arthritis highly likely. However, the patient is not always able to differentiate between swelling arising from the joint itself and periarticular structures. In the case of periodic arthritis syndrome, it is important for the physician to communicate to the patient the importance of an examination when their joints are swollen.

Joint Stiffness

Stiffness occurs in joints that are inflamed. Therefore, stiffness is a feature of inflammatory arthritis irrespective of the cause of the inflammation. In osteoarthritis, there is often severe synovitis, which histologically may be indistinguishable from rheumatoid arthritis. Consequently, osteoarthritis often involves stiffness in the joints. Joint stiffness is particularly severe in the morning due to the nocturnal accumulation of body fluid. This stiffness characteristically diminishes during the day when the patient exercises. Stiffness, presumably due to fluid accumulation in inflamed tissue, is often noticeable after rest and is referred to as "jelling."

Joint Locking, Snapping, and Instability

Locking of a joint indicates a loose body in the joint. The most common example is in meniscal lesions of the knee. Locking may be caused by joint "mice" consisting of broken-off cartilage or bone, as in osteoarthritis, osteochondromatosis, and osteochondritis dessicans. Joint locking may also occur with joint instability when the joint dislocates.

Patients with a generalized hypermobility syndrome may experience a snapping sensation with movement of the scapulothoracic joint. The medial edge of the scapula is often tender and pain may be relieved with local corticosteroid injections. Radiographs may detect an underlying bone abnormality, such as an osteoid osteoma or a spur on the scapular border. A snapping hip is produced when the fascia lata slides over the greater trochanter of the femur, especially on flexion or rotation of the hip.

Patients frequently complain of joints, the knee in particular, giving way or buckling. This can be caused by a number of conditions affecting the joint, such as a torn meniscus, loose body, arthritis, and weakness of the vastus medialus.

MUSCLE WEAKNESS

Weakness of muscles is always an important symptom. It can be due to a variety of conditions affecting muscles either primarily or neurologically. It is important to determine the character, timing, and location of the weakness. Weakness may be associated with inflammatory arthritis due to disuse atrophy. Progressive generalized muscle weakness during the day should suggest myasthenia gravis. Shoulder and hip girdle weakness is common in myopathy. Peripheral muscle weakness is more common in neurological disorders.

FUNCTION

It is important to determine the degree of physical impairment from which a patient is suffering. When describing the arthritic patient, we prefer to use the term locomotor failure, which is analogous to commonly used terms such as heart, respiratory, or renal failure. In patients with severe crippling arthritis, the quality of life is often greatly impaired. As yet, there is no standard method for measuring the quality of life. Interested readers are referred to the reviews by Liang, et al and Bellamy (2,3).

Movements that cause an increase in pain are important when taking a history of the musculoskeletal system. For example, a patient experiencing knee pain when descending stairs should be suspected of having patellofemoral disease. A patient with lumbar pain and bilateral sciatic pain that is relieved when holding the spine in flexion should be suspected of having spinal stenosis. Grasping may increase the pain associated with epicondylitis and carrying a heavy object may aggravate a thoracic outlet syndrome. Pain and functional impairment are often associated in rheumatic patients. Therefore, it is important to determine whether loss of function is due to pain or other causes, such as muscle weakness or joint deformity.

FEVER

Fever is not a common finding in disease of the musculoskeletal system. Although fever should immediately suggest infection, patients with rheumatoid arthritis, systemic lupus erythematosus, and other connective tissue diseases may also be febrile during acute exacerbations. Fever associated with juvenile rheumatoid arthritis is often low or absent in the early part of the day and rises to high levels in the afternoon and evening (Fig. 2–2). The fever is sim-

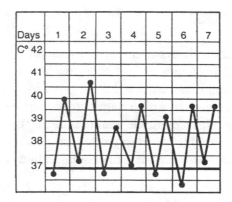

Figure 2–2. Typical temperature in Still's disease. Note how the temperature is often normal in the morning and elevated in the evening.

ilar to that associated with acute bacterial infections. It is often accompanied by the typical rash of juvenile rheumatoid arthritis. Rheumatic fever usually presents with a continuous fever (Fig. 2–3), which rarely reaches high temperatures.

Hyperpyrexia is currently of historical interest in developed countries. Fever may accompany inflammatory arthritis and may be short-lived or prolonged. Recently, we saw a patient with a prolonged low-grade fever due to chronic gouty arthritis. It is important for the clinician to realize that a septic arthritis may not be accompanied by fever. This is particularly true in rheumatoid arthritis. Chills and severe shivering or rigors may indicate bacteriemia or viremia. Night sweats are common in low-grade infections, such as brucellosis and tuberculosis.

A febrile incident may precede arthritis and may be due to rubella, which is sometimes associated with a polyarthritis in the adult. Therefore, a history of fever is important, even if the patient is afebrile at the time of consultation.

Drug History

Iatrogenic illness is common in all medical specialties. Antirheumatic drugs produce approximately half of the reported side effects sent to the Committee of Safety of Medicines in the United Kingdom. Phenylbutazone and oxyphenbutazone are the most common drug causes of aplastic anemia. When expressed as a percentage of the number of prescriptions, gold is the most toxic compound in the British Pharmacopoeia. Therefore, it is essential

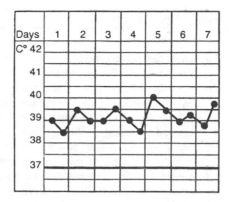

Figure 2–3. Continuous fever in acute rheumatic fever.

to obtain information regarding what drugs the patient has received and whether there are any known drug allergies. The most common complications of antirheumatic drug therapy are gastrointestinal. Although dyspepsia is the most frequent problem, acute and chronic blood loss are more serious complications. Iatrogenic causes of musculoskeletal symptoms have recently been reviewed (4).

A number of drugs may cause connective tissue disease syndromes. Also, acute gouty arthritis may complicate several drugs (Table 2–1).

An illness resembling systemic lupus erythematosus may occur during treatment with procainamide, hydralazine, anticonvulsants, isonicotinic acid hydrazide, and other miscellaneous drugs. The most common cause is procainamide. Approximately half of the patients taking this drug develop antinuclear factors. Hydralazine also commonly causes systemic lupus-like disease. Despite approximately 20% of patients who develop antinuclear factors when using isonicotinic acid hydrazide, this drug has rarely been implicated as a cause of illness resembling systemic lupus. Therefore, the finding of a positive antinuclear factor does not indicate a diagnosis of systemic lupus erythematosus and is not a contraindication to continued therapy.

There are important clinical differences between spontaneous systemic lupus erythematosus and drug-induced disease. For example, the sex ratio is approximately equal in the drug-induced disease. Also, blacks are less commonly affected than whites by the drug-induced disease. In addition, central nervous system and renal complications are less common in the drug-induced disease and the condition responds to withdrawal of the drug. The

Table 2–1. Drugs Causing Hyperuricemia

Drugs	Comments
Salicylates	In low doses (large doses, such as 3g to 5g/day increase urinary excretion of uric acid, with resultant hyperuricemia).
Thiazide diuretics	Approximately 40%–50% of patients treated with these drugs develop hyperuricemia. Potassium reduces elevated serum urate levels in patients treated with chlorothiazide, but requires confirmation. Probenecid reverses the hyperuricemia, which can also be controlled by allopurinol.
Other diuretics	Chlorthalidone, ethacrynic acid, and furosemide.
Ethanol	Acutely intoxicated patients develop hyperuricemia.
Acetazolamide	Causes slight increase in serum uric acid concentration.
Aluminum nicotinate	Nothing
Angiotensin	Produces hyperuricemia due to efferent vasoconstriction in the kidney.
Cyclosporine	Produces hyperuricemia by affecting renal function.
Diazoxide	Causes decrease in urinary uric acid excretion.
Epinephrine and nonepinephrine	Causes hyperuricemia by efferent vasoconstriction in the kidney.
Ethambutol	Decreases urinary excretion of uric acid, which is not reversed by small doses of salicylates.
Gentamicin	Causes hyperuricemia, but requires confirmation.
Laxative abuse	Nothing.
Levodopa	Nothing.
Mecamylamine	Nothing.
Methotrexate	Acute gout has been reported after intravenous administration in psoriatic arthritis.
Methoxyflurane	Nothing.
Nicotinic acid	Hyperuricemia occurs with large doses of 3g/day or more.
Phenothiazines	Hyperuricemia has been reported, but requires confirmation.
Pyrazinamide	Inhibits the excretion of uric acid, which can be prevented by concomitant aspirin, aminosalicylic acid, and phenylbutazone therapy.
Quinethazone Trimterene	Reported to cause hyperuricemia in 20% of patients, but not confirmed by some authors. Hyperuricemia augmented by hydrochlorothiazide.

laboratory differences between spontaneous and drug-induced diseases are discussed in Chapter 5.

Exposure to vinyl chloride may cause Raynaud's phenomenon and acroosteolysis similar to progressive systemic sclerosis. Bleomycin can cause pulmonary fibrosis. Methysergide and hydralazine have been reported to cause retroperitoneal fibrosis. Chronic ingestion of ethyl alcohol is associated with Dupuytren's contracture.

Any therapeutic substance that causes serum sickness may result in an inflammatory polyarthritis. Corticosteroid therapy, when given in excess, may produce an exacerbation of arthritis. This is due to acute muscle atrophy, which gives rise to pseudorheumatism and is seen in Cushing's syndrome due to nontherapeutic causes. Fractures of bones may occur as a result of osteoporosis induced by long-term corticosteroid therapy and may produce severe pain.

Drug interactions may also be the source of symptoms. The clinically established interactions with antirheumatic drugs are summarized in Table 2–2 (5–7).

Surgical treatment may cause hyperuricemia or provoke an attack of acute gout. Patients undergoing dialysis or transplantation for renal disease may develop a variety of musculoskeletal complications, including arthritis. The ileal bypass operation for obesity, which is rarely performed, may result in a polyarthritis. Removal of the affected segment in ulcerative colitis may cause remission of the arthritis associated with this disease.

FAMILY HISTORY

A family history is important in diseases with a strong genetic basis (e.g., hemophilia, ochronosis, and sickle-cell disease). However, the absence of a family history does not exclude the risk of a genetically determined disease. For example, one-third of patients with hemophilia have no family members affected by the disease.

A number of diseases, such as ankylosing spondylitis and gout, have an increased number of family members affected. The precise pattern of inheritance is not always clear. In ankylosing spondylitis, the genetics have recently been explained on the basis of a close association with the tissue type HLA-B27. In rheumatoid arthritis, toxicity to chrysotherapy is associated with the HLA-DR3 haplotype, which is associated with more severe diseases. In gout, there have been a number of enzyme defects isolated. In the future, the family inheritance of gout will be determined by assay of enzymes of purine biosynthesis.

A family history does not necessarily imply a genetic basis. Many diseases seen within families (e.g., syphylis and tuberculosis) are due to environmental causes. In clinical practice, a family history is not usually of any value in the diagnosis or management of the disease.

Table 2–2. Clinically Established Interactions with Antirheumatic Drugs

		Clinical Effect	Mechanism of Interaction
Allopurinol	Azathioprine	Increased azathioprine toxicity	Decreased azathioprine metabolism
	Cyclophosphamide	Increased cyclophosphamide toxicity	Not determined
	Mercaptopurine	Increased mercaptopurine toxicity	Decreased mercaptopurine metabolism
	Oral anticoagulants	Increased anticoagulant activity	Inhibition of microsomal enzymes
Aspirin and Salicylates	Acetazolamide	Increased CNS toxicity of acetazolamide	Increased plasma nonionized salicylate with increased brain concentrations
	Alcohol	Erosive gastritis	Additive
	Antacids	Decreased plasma salicylate concentrations	Increased renal clearance of salicylate
	Aminosalicylic acid (PAS)	Decreased salicylate levels	Increased renal clearance
	Bumetanide	Decreased diuretic effect	Inhibition of prostaglandin synthesis
	Captopril	Decreased antihypertensive effect	Possible inhibition of prostaglandin
	Dipyridamole	Increased effects on function	Synergism
	Heparin	Potential increased bleeding risk, especially with aspirin	Inhibition of platelet function. Theoretical rather than proven risk
	Methotrexate	Increased methotrexate toxicity	Displacement of methotrexate from protein binding sites and decreased renal excretion of methotrexate. Latter probably more serious effect
	Moxalactam	Increased bleeding risk	Additive effect, but risk is only potential
	Oral anticoagulants	Increased bleeding risk	Inhibition of platelet function with aspirin additive effect. With high doses of salicylates increased hypoprothrombinemic effect
	Oral hypoglycemics	Increased hypoglycemia	Additive effect with displacement of hypoglycemics from protein binding sites, especially chlorpropamide

	Probenecid	Decreased uricosuric effect	Mechanism not established, but important in the treatment of gout
	Sulphinpyrazone	Decreased uricosuric effect	Mechanism not established, but important in the treatment of gout
Corticosteroids	Barbiturates	Decreased corticosteroid effect	Induction of microsomal enzymes
	Diuretics	Increased loss of urinary potassium	Additive. Note: Not with spironolactone and triamterene
	Ephedrine	Decreased dexamethasone effect	Not yet determined
	Phenytoin	Decreased corticosteroid effect	Induction of microsomal enzymes
	Rifampin	Decreased corticosteroid effect	Induction of microsomal enzymes
	Salicylates	Increased salicylate toxicity	A clinical problem due to the fact that corticosteroids induce salicylate metabolism. When oral corticosteroids are being reduced in a patient on chronic salicylate medication the serum salicylate concentration may rise to toxic levels.
Indomethacin	Antacids	Decreased indomethacin effect	Decreased indomethacin absorption
	Beta-adrenergic blockers	Decreased antihypertensive effect	Possibly by prostaglandin inhibition
	Diuretics	Decreased antihypertensive and natriuretic effect	Possibly by prostaglandin inhibition. Seen with thiazides and furosemide
	Lithium	Increased lithium toxicity	Decreased lithium renal excretion
	Oral anticoagulants	Potential increased bleeding risk	Inhibition of platelet function
	Sympathomimetics	Hypertension and hypertensive crisis	Inhibition of norepinephrine uptake
Methotrexate	Phenylbutazone and oxyphenbutazone	Increased methotrexate effect	Not established
	Probenecid	Increased methotrexate effect	Decreased methotrexate renal excretion
	Salicylates	Increased methotrexate effect	Displacement of methotrexate from protein binding sites and diminished renal excretion of methotrexate. Latter probably most serious effect

Table 2-2. Clinically Established Interactions with Antirheumatic Drugs—*continued*

		Clinical Effect	Mechanism of Interaction
Naproxen	Probenecid	Increased naproxen effects	Probenecid inhibits hepatic metabolism of naproxen
Penicillamine	Digoxin	Decreased digoxin effect	Not established
	Oral iron	Decreased absorption of both drugs	Both drugs actively absorbed from gastrointestinal tract and interfere with each other's absorption
Phenylbutazone and Oxyphenbutazone	Oral anticoagulants	Increased anticoagulant effect	Displacement from protein binding sites and inhibition of microsomal enzymes
	Oral hypoglycemics	Increased sulphonylurea hypoglycemia	Inhibition of microsomal enzymes
	Methotrexate	Increased methotrexate effect	Not established
	Phenytoin	Increased phenytoin toxicity	Inhibition of microsomal enzymes
Probenecid	Aminosalicylic acid (PAS)	Increased PAS toxicity	Inhibition of renal excretion
	Bumetanide	Possible decreased diuretic effect	Decreased renal excretion of bumetanide. May require increased bumetanide dosage
	Methotrexate	Increased methotrexate effect	Decreased renal excretion of methotrexate
	Salicylates	Decreased uricosuric effect	Not established

SOCIAL AND OCCUPATIONAL HISTORY

Patients who are severely crippled are often confined to their homes. Therefore, it is important to question patients about circumstances at home. Relevant questions include the accessibility of a toilet, whether there are stairs to climb, availability of bathing facilities and the ability to use them, can they do their own messages, can they cook or do they require a service, do friends and relatives visit regularly, and is help easily obtained? Inquiries should also be made regarding chiropody, physiotherapy, and occupational therapy. Depending on the answers to these questions, it may be necessary to have the patient moved to another house. The help of a medical social worker is often invaluable.

The patient's occupation may be relevant to musculoskeletal symptomatology (8). For example, North Sea divers and tunnellers may suffer severe bone pain due to Caisson's disease. Miners frequently suffer lumbar disc degeneration and osteoarthritis of the knees. The use of vibratory tools may cause Raynaud's phenomenon and avascular necrosis of the lunate. Veterinary surgeons and dairymen are likely to develop brucellosis.

In addition to these causal relationships between occupation and disease of the locomotor system, the patient's occupation may be important for other reasons. For instance, a long-distance truck driver who has to unload heavy material will be unable to perform the job if suffering from rheumatoid arthritis that is becoming progressively worse. A sedentary occupation that involves bending over a desk (e.g., a draftsman) is difficult if the patient suffers from ankylosing spondylitis.

It is also important to consider the patient's life-style. For example, a heavy whiskey drinker with gouty arthritis is unlikely to comply with instructions regarding drug therapy. A promiscuous young man with a seronegative polyarthritis affecting the knees and ankles should immediately be suspected of having Reiter's disease. In a young woman who is promiscuous, gonococcal arthritis should be excluded if another diagnosis is not readily established.

In rheumatic diseases, it is important to assess the patient's attitude about the disease and the consequent functional disability. Patients with rheumatoid arthritis are often depressed, especially when they suffer functional disability and become dependent on others. It is important to determine the patient's attitude before performing operations on the joints. These operations frequently require the cooperation of the patient in physiotherapy, which may be painful.

Taylor referred to the "predicament" of illness (e.g., social, psychologic, and economic implications) (9). It would not be surprising to find that he had the arthritis sufferer in mind, because no physical illness produces such a predicament over such a long time.

REFERENCES

1. Wyke B. The neurology of joints: a review of general principles. Clin Rheum Dis 1981;1:223–239.
2. Liang MH, Cullen KE, Larson MG. Measuring function and health status in rheumatic disease clinical trials. Clin Rheum Dis 1983;9:531–539.
3. Bellamy N. Musculoskeletal clinical metrology. Dordrecht: Kluwer Academic Publishers, 1991.
4. Kahn MF, ed. Drug induced rheumatic diseases. In: Bailliere's Clinical Rheumatology, Vol 5. London: Bailliere Tindall, 1991.
5. Anonymous. Adverse interactions of drugs. Medical Letter 1981;23:17–28 and 1984;26: 11–14.
6. Rizack, MA. The medical letter handbook of adverse drug interactions. New York: The Medical Letter, Inc., 1995.
7. Brooks PM. Drug modification of inflammation—nonsteroidal antiinflammatory drugs. In: Maddison PJ, Isenberg DA, Woo P, et al, eds. Oxford textbook of rheumatology. Vol 1. Oxford, England: Oxford Medical Publishers, 1993.
8. Balint GP, Buchanan WW, eds. Occupational rheumatic diseases. In: Balliere's Clinical Rheumatology, Vol 3. London: Bailliere Tindall, 1989.
9. Taylor DC. The components of sickness: diseases, illnesses and predicaments. Lancet 1971;2:1008.

Regional Examination of the Limbs and Spine | 3

SHOULDER

Shoulder pain is common. In the United Kingdom, it has been estimated that 1 in 170 of the adult population consult their family physician each year for shoulder pain (1). Shoulder pain may be caused by diseases affecting the structures around the shoulder and systemic illnesses. For example, pain in the shoulder may be from disease of the cervical spine, gallbladder, spleen, diaphragm, myocardium, or apex of the lung (e.g., Pancoast tumor).

The shoulder joint is one of the most common causes of regional pain affecting the musculoskeletal system and, anatomically, one of the most complex articulations in the body. There are four components of the shoulder. These are the glenhumeral joint, acromioclavicular joint, sternoclavicular joint, and scapulothoracic joint.

The glenhumeral joint is a ball and socket joint, which is secured by muscles rather than bones or ligaments. The depth of the glenoid cavity is increased by a ring of fibrocartilage called the labrum, which is attached to the rim of the cavity. On the inferior aspect of the glenhumeral joint, the capsule and synovium form a pouch-like axillary recess, which allows for a large range of movement. The long head of biceps originates at the apex of the glenoid cavity and is enclosed in a sheath of synovial membrane. It passes over the head of the humerus and descends in the bicipital groove (intertubercular sulcus). The acromioclavicular and sternoclavicular joints depend on ligaments for stability. Although the scapulothoracic articulation is not a true joint, it functions as an integral part of the shoulder complex.

The rotator cuff consists of four strap-like muscles; the subscapularis, supraspinatus, infraspinatus, and teres minor. The tendons merge to form a cuff, which blends with the lateral aspect of the capsule of the glenhumeral joint as it passes over the head of the humerus and under the subacromial bursa to be inserted into the greater (supraspinatus, infraspinatus, and teres minor) and lesser (subscapularis) tuberosities of the humerus. The supraspinatus is the most important of these muscles because it fixes the head of the humerus in the glenoid cavity when the deltoid muscle abducts the shoulder.

The muscles of the rotator cuff function in a confined space. Therefore, they are subject to attrition and trauma. Degeneration, with thinning and rupture of the rotator cuff, is also due to peculiarities in the microvascular supply. There is a "critical zone" of relative avascularity in the supraspinatus tendon (Fig. 3–1) that predisposes elderly patients to degeneration, basic calcium phosphate deposition, and rupture of the tendon. A rotator cuff tear results in subluxation in a superior direction of the humeral head, although the reverse might be expected. The acromiohumeral interval, the distance between the acromion process and the humeral head, is important when diagnosing the complication of impingement syndrome. The acromiohumeral interval is usu-

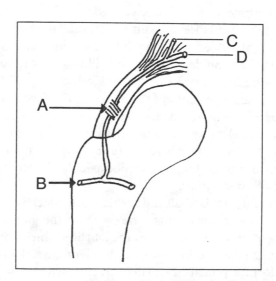

Figure 3–1. The critical zone (a) in the rotator cuff is situated where the branch from the anterior circumflex artery (b) entering from the humerus below meets the artery formed by the suprascapular (c) and subscapular (d) branches entering from the muscle above. The critical zone is rendered relatively ischemic when the tendon passes over the humeral head. (Adapted from: Cailliet R. Shoulder pain, 2nd ed. Philadelphia: F.A. Davis Co., 1981.)

ally 7–14 mm and can be measured on straight radiographs. Cone and colleagues provide an excellent discussion of radiological features (2).

The shoulder can move in many directions. The most important directions include forward flexion (180°), extension (50°), abduction (lateral elevation at 180°), adduction (45°), and external and internal rotation (90°). These movements are illustrated in Figure 3–2.

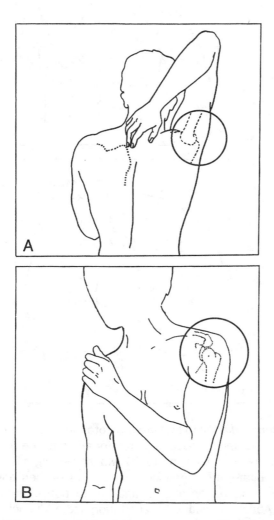

Figure 3–2. Active range of shoulder movements. (A) to test for abduction and external rotation the patient reaches behind his head and touches the superior angle of the opposite scapula; (B) to test for internal rotation and adduction the patient touches the opposite acromion (Apley's scratch test).

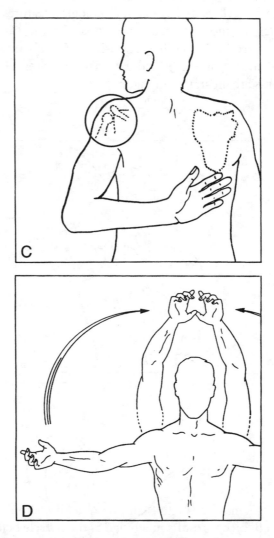

Figure 3–2—*continued.* (C) to test internal rotation and adduction the patient reaches behind his back to touch the inferior angle of the opposite scapula; (D) the patient abducts his arms to 90° while keeping his elbows straight, which provides bilateral comparison. The patient is then asked to turn his palms upward in supination and continue abduction until his hands touch over his head. Abduction and external rotation and adduction and internal rotation can also be done with both arms simultaneously, thus facilitating examination for symmetry of motion and small losses of movement.

Examination of the shoulder should include conventional inspection, palpation, and examination of passive and active movements.

Inspection

For accurate inspection, the patient must be undressed. The shoulders should be visually examined from the front, back, and side. Cutaneous rashes, such as herpes zoster and the rash of dermatomyositis, may immediately indicate shoulder symptoms. The shoulders should also be examined for obvious differences in level, which may be due to anatomical deformities, such as scoliosis, fracture of the clavicle, and Sprengel's deformity. An unequal level of the shoulders may also be the result of functional abnormalities, such as spasticity of the cervical muscles and trapezius, resulting from pain in the cervical spine or thoracic outlet, or weakness of muscles due to a nerve lesion. The level of the shoulder is usually lower on the dominant side, especially in athletes.

Atrophy of muscles around the shoulder, if bilateral, is most likely due to a myopathy, polymyositis, motor neuron disease, or corticosteroid-induced myopathy. When atrophy is unilateral, a local cause is more likely, such as a root or peripheral nerve lesion. Deltoid muscle wasting occurs with paralysis of the axillary (circumflex) nerve (C5). Wasting of the trapezius is due to a lesion of the accessory nerve and branches of the third and fourth cervical nerves. A lesion of the suprascapular nerve (C5 and C6) results in wasting of the supraspinatus and infraspinatus muscles. A long thoracic nerve lesion (C5, C6, and C7) causes wasting of the serratus anterior muscle and "winging" of the scapula. A lesion involving cervical nerve roots causes more widespread muscle wasting. Muscle wasting may also be due to arthritis affecting the shoulder joint and inactivity after a traumatic injury.

In the past, tuberculosis of the shoulder joint caused marked wasting of muscle. Currently, rheumatoid arthritis is the most common disease causing severe joint involvement at the shoulder. Swelling on the lateral aspect of the shoulder is rare and may be due to effusion in the subdeltoid bursa. Rupture of the long head of biceps results in a bulge of the muscle belly when the elbow is flexed against resistance. A step deformity or sulcus sign may be evident below the acromion process when the acromioclavicular joint is dislocated. Shoulder joint effusions are usually caused by rheumatoid arthritis. Because the effusions usually bulge forward they can be readily palpated.

Palpation

When examining the shoulder, it is important to identify the normal anatomical landmarks, such as the acromion, coracoid process (two fingerbreadths medially and two fingerbreadths below the lateral end of the clavicle), glenohumeral joint space, acromioclavicular and sternoclavicular joints, greater tubercle of the humerus, and spine of the scapula. Arthritis affecting the glenohumeral joint results in tenderness when pressure is applied along the joint margin. Arthritis affecting the acromioclavicular and sternoclavicular joints also results in tenderness when these joints are subjected to pressure.

Pain and tenderness in the acromioclavicular and sternoclavicular joints tends to be localized in the joints. Pain in the shoulder may be experienced over the C5 dermatome, but seldom extends below the elbow. The biceps tendon may be tender on palpation, indicating bicipital tendinitis. Tenosynovitis of the long head of biceps may result from repetitive trauma and wide or deep bicipital grooves. Wide bicipital grooves cause undue lateral movement of the tendon, while deep grooves are too narrow, causing compression.

The supraspinatus tendon and subdeltoid bursa are frequent sources of pain in the shoulder. Movement of the shoulder and palpation are necessary to differentiate supraspinatus tendinitis and subdeltoid bursitis. Although abduction causes pain in both conditions, resisted abduction only causes pain in supraspinatus tendinitis. In their excellent book on practical orthopaedic medicine, Corrigan and Maitland claim that a gap may be felt in the rupture of the supraspinatus tendon, but we have not been able to confirm this statement (3). Patients with diabetes mellitus are particularly prone to supraspinatus tendinitis. This is associated with patients with carpal tunnel syndrome, Dupuytren's contracture, flexor tenosynovitis, and sclerodactyly, all which may result from nonreversible glycosylation of collagen crosslinks.

Crepitus may be felt with movement of an arthritic shoulder. A painful click suggests a possible labrum tear or subluxing bicipital tendon. A clump indicates major instability.

Finally, it is important to examine the armpit for conditions such as lymphadenopathy or a brachial artery aneurysm. Patients who have had a myocardial infarction, hemiplegia, or herpes zoster may develop pain and stiffness in the shoulder associated with vasomotor disturbances in the corresponding hand. This "shoulder-hand" syndrome is resistant to treatment and is probably due to a reflex sympathetic disturbance. Radiograph changes include osteoporosis. A radionuclide bone scan shows an increased uptake in the involved hand.

In rheumatoid arthritis, Baker's cysts may develop, which are similar to those behind the knee. These cysts may extend down the medial aspect of the arm as far as the elbow. Also, fibrositis nodules may be palpated along the spine of the trapezium. Multiple maneuvres have been described to identify causes of shoulder pain. The interested reader should consult the book by

Corrigan and Maitland, mentioned previously, and the texts of Hoppenfeld, Little, and Magee (4,5,6).

Examination of Movement

Movement of the shoulder should be active and passive. These movements should be examined separately. Active movements test the power of different muscles, whereas passive movements test joint function and associated periarticular joint tissues. When active movements are tested against resistance, muscle weakness is apparent and pain is elicited when tendinitis and enthesitis (e.g., where tendon is inserted into periosteum or bone) are present. These different types of movement may seem tedious and time-consuming, but probably are no more so than unnecessary, expensive laboratory investigations.

The patient is best examined stripped to the waist and sitting upright in a chair. Active, active-resisted, and passive movements should be examined. The examining physician should mark the inferior tip of the scapula to ascertain when scapular movement occurs (Fig. 3–3). If movement of the scapula occurs immediately on abduction, this indicates that the glenohumeral joint is not moving properly. The scapula should only begin to move after 20° abduction. The clinical methods that should be employed in movement of the shoulder are illustrated in Figure 3–4.

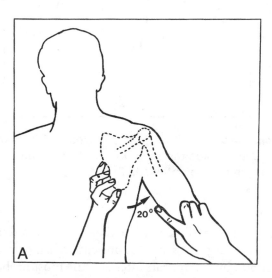

Figure 3–3.   Passive range of abduction of the shoulder. (A) The scapula will not move until the arm is abducted 20°. Thereafter, the glenohumeral joint and scapulothoracic articulation move in a ratio of 2:1.

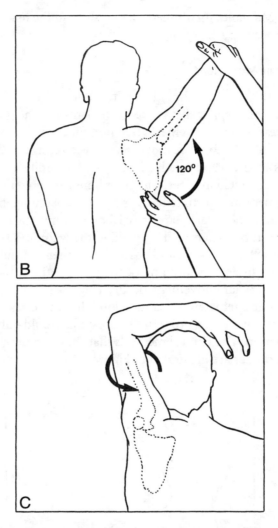

Figure 3–3—*continued.* (B) Abduction continues to 120°, when the surgical neck of the humerus strikes the acromion. (C) If the humerus is externally rotated, full abduction can be obtained.

Resisted supination of the arm when the elbow is flexed causes pain when bicipital tendinitis (Yergason's sign) is present. When pain occurs on active abduction after 30°, but disappears after 70°, the most probable diagnosis is supraspinatus tendinitis. This is the "middle arc syndrome" and occurs when the inflamed and swollen tendon is constricted under the acromion process and the coracohumeral ligament between 30° and 70° abduction. If the pa-

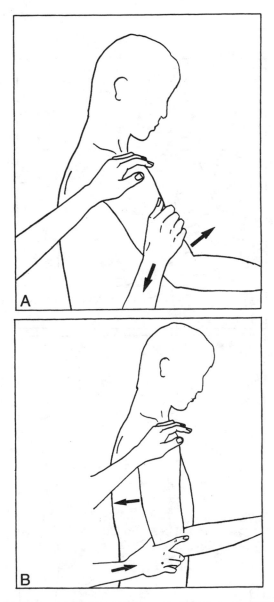

Figure 3–4. Muscle testing in the shoulder. (A) Flexion; (B) Extension; (C) Abduction; (D) Adduction; (E) External rotation with elbow against body; (F) Internal rotation with elbow against body; (G) Scapular elevation; (H) Scapular retraction; (I) Scapular protraction. Winging of the scapula indicates weakness of serratus anterior.

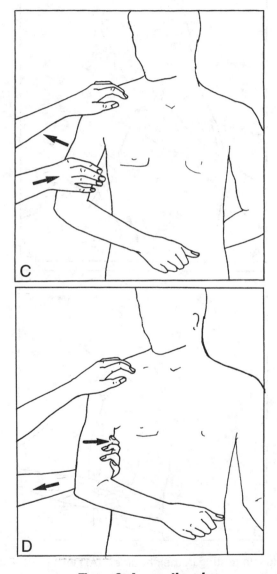

Figure 3–4—*continued.*

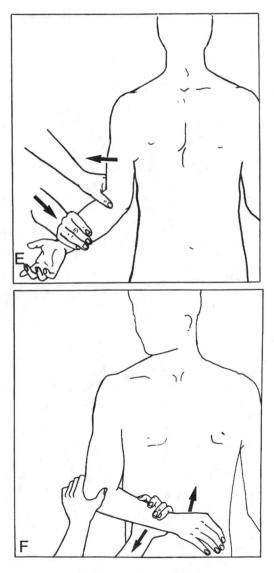

Figure 3–4—continued.

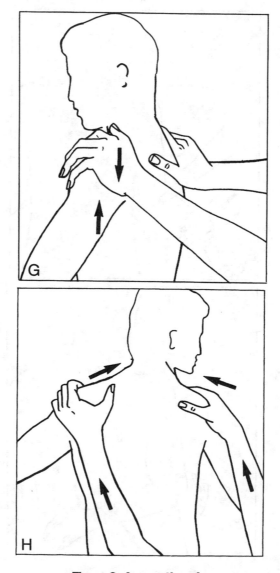

Figure 3–4—*continued.*

Figure 3–4—continued.

tient experiences pain on full elevation (80°), it is probably due to disease of the acromioclavicular joint. When supraspinatus tendinitis or rupture of the supraspinatus tendon is present, the arm suddenly drops from the abduction position. This is called the "drop-sign" (Fig. 3–5). When the patient is unable to perform abduction for the first 30°, this is usually due to supraspinatus tendon rupture or weakness. The patient often overcomes the inability to perform abduction for the first 30° by maneuvering the body so that the deltoid muscle performs abduction beyond the first 30° (Fig. 3–6).

Injection

The shoulder joint may be aspirated by an anterior or posterior approach (Fig. 3–7). The anterior joint line can be palpated 0.5–1.0 cm lateral to the tip of the coracoid process. This approach is difficult and may result in injection of the bicipital tendon, causing rupture. In the posterior approach, the injection is made horizontally 2 cm below and 2 cm medially from the base of the acromion. This approach is recommended for those who are not experienced. The lateral approach is used for treating supraspinatus tendinitis. The injection is made below the acromion process after the needle has been advanced 2.5 cm. A number 23 needle should be used when injecting corticosteroid. We use triamcinolone hexacetonide (Kenalog), which has a long half

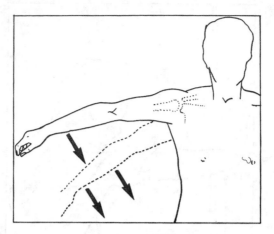

Figure 3–5.   The drop arm test. The patient fully abducts his arm and then slowly lowers it to his side. If there is a tear in the rotator cuff, the patient's arm will drop as he attempts to lower it. If he is able to hold his arm in abduction, a gentle push on the forearm will cause the arm to drop to his side.

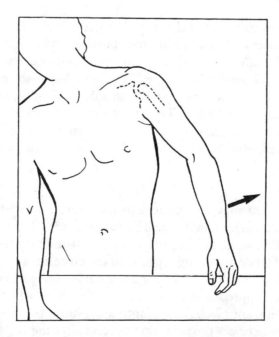

Figure 3–6.   Difficulty in initiating abduction is suggestive of a shoulder cuff lesion or supraspinatus tendon tear.

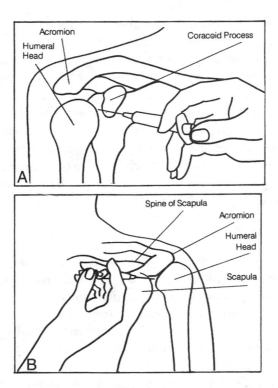

Figure 3–7. (A) Technique of intraarticular injection of the shoulder (anterior approach); (B) Technique of intraarticular injection of the shoulder (posterior approach).

life, for intraarticular injection, and methylprednisone (Medrol) for injection of soft tissue lesions. For shoulder injections, a dose of 40 mg of triamcinolone hexacetonide usually suffices. For supraspinatus tendinitis, an injection of 10 mg of methylprednisone with 2–3 ml of a local anesthetic, such as lidocaine, in a 1% solution is usually sufficient.

ELBOW

Although the elbow is a much more simple joint than the shoulder, it also consists of several articulations. These include the radiohumeral joint, ulnohumeral joint, and superior or proximal radioulnar joint.

These three joints comprise a single compound joint. The capsule and joint cavity are continuous for all three articulations of the elbow. Therefore, injury or pathology of one component also affects the other components. The radial head articulates with the capitulum and the ulna with the trochlea. The ulnohumeral joint allows flexion and extension, while the radiohumeral and superior radioulnar joints permit rotation. The carrying angle is 5–10° in males and 10–15° in females. If the carrying angle is greater than 15°, cubitus valgus is present. This is classically seen in gonadal dysgenesis (chromsome X or Turner's syndrome). Cubitus valgus and varus, often referred to as "gunstock" deformity because of their shape, frequently result from fractures around the elbow. Subtle fractures in children may be suspected if the normal carrying angle exceeds 5° (Bowman's angle) between the injured and normal elbows, as defined radiologically.

Inspection

Inspection of the elbow may reveal traumatic lesions and scars, swelling of the joint, olecranon bursitis, rheumatoid subcutaneous nodules, flexion contraction of the joint, and varus or valgus deformity of the joint. Swelling of the joint due to hypertrophied synovium appears as a mass over the radial head anteriorly and between the olecranon and lateral epicondyle, the paraolecranon space. The synovial membrane lies close to the surface of this space. When the elbow is fully extended, the two epicondyles and olecranon process should form a straight line, but become an isosceles triangle when the elbow is flexed (triangle sign).

Palpation

Palpation of the anatomic landmarks should be carried out in a systematic fashion. The epicondyles can be easily palpated laterally and medially. When tennis elbow (lateral epicondyle) or golfer's elbow (medial epicondyle) are present, the epicondyles are tender on palpation. The extensor muscles of the forearm originate on the lateral epicondyle of the humerus, whereas the flexor muscles originate on the medial condyle. The groove for the ulnar nerve, which lies just posterior to the medial epicondyle, can be palpated. Hypertrophy or a neuroma of the ulnar nerve can easily be recognized.

An injury to the area of the cubital tunnel, through which the ulnar nerve passes, may lead to problems causing pressure on the ulnar nerve (tardy ulnar palsy). The ulnar nerve may sublux in and out of its groove. Tinel's sign, elicited by tapping over the ulnar nerve, produces tingling in the ulnar distribution of the nerve in the forearm and hand. Tingling may also be elicited when the elbow is held in full flexion for 5 minutes, indicating a cubital tunnel syndrome. Diagnosis of ulnar nerve compression in the cubital tunnel can be confirmed with nerve conduction studies, which also exclude compression of the C8 nerve root in the neck as a possible diagnosis. An entrapment neuropathy of the medial nerve may occur under a ligament of Struthers or between the two heads of the pronator teres muscle. Symptoms are similar to those of carpal tunnel syndrome in the wrist. If entrapment is between the two heads of the pronator teres muscle, symptoms will be aggravated by pronation of the forearm.

In inflammatory joint disease and lateral epicondylitis, the radial head is often tender when pressure is applied. The joint margin can be palpated on both sides of the olecranon. Synovial hypertrophy can also be palpated in this region. The ulnar border of the olecranon should be palpated for rheumatoid subcutaneous nodules, which may be multiple and 3–4 cm in size. The nodules usually feel firm and rubbery and may develop a necrotic center, which discharges and may become infected. Patients with rheumatoid subcutaneous nodules are usually seropositive for IgM rheumatoid factor, which is often in high titre, and susceptible to other systemic manifestations. Gouty tophi also develop at the ulnar border of the olecranon and may be mistaken for rheumatoid nodules. If the discharge is chalky and contains uric acid crystals, gouty tophi are present. Olecranon bursitis is usually the result of repeated trauma and is common in miners ("beat elbow"). The inflamed bursa is usually soft and fluctuant, perhaps why it is often described as a "goose egg." Rheumatoid arthritis and gouty arthritis can cause olecranon bursitis. In acute attacks of gout, the leucocyte count is never as high as the count commonly found in joints (2000 cells/mm^3).

Xanthomata also occur over the ulnar border of the olecranon. In the antecubital fossa, the relationships between the biceps tendon, brachial artery, and median nerves provide useful anatomical knowledge (Fig. 3–8). Injury to the brachial artery in the antecubital fossa may result in Volkmann's ischemic contracture. Entrapment of the radial nerve at the elbow may cause sensory symptoms if the superficial branch is damaged. The posterior interosseous nerve, the major branch, is more likely to be compressed between the two heads of the supinator muscle (arcade of Frohse) and leads to symptoms similar to medial epicondylitis, wasting of the extensor muscles of the forearm, and weakness of the wrist and fingers.

In inflammatory joint disease affecting the elbow, such as rheumatoid arthritis, rupture of the joint may occur. This is characterized by acute pain and swelling of the forearm and hand. The diagnosis can be confirmed by an arthrogram. In toddlers, dislocation of the head of the radius may occur with a tug of the arm. This "pulled elbow" lesion, sometimes referred to as "nursemaid's elbow," was originally thought to be due to the small radial head, which could be pulled through the annular ligament. However, the radial head in toddlers is larger than the neck and cannot be pulled through the an-

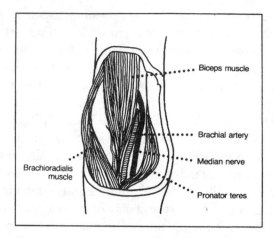

Figure 3–8. Anatomical relations of the antecubital fossa. The fossa's base is formed by a line drawn between the two epicondyles of the humerus. The lateral boundary is formed by the brachioradialis and the medial boundary by the pronator teres. The structures in the fossa from lateral to medial are biceps tendon, brachial artery, and median nerve.

nular ligament. The annular ligament is weak and thin in toddlers and can be ruptured by a sudden upward pull on the arm.

Movement

The elbow joint should be examined with the patient in the sitting position. Active, active-resisted, and passive movement should be tested. It is important to "fix" the elbow when testing movement (Fig. 3–9). Restricted pronation and supination of the elbow may be entirely due to distal radioulnar joint disease. This is because the elbow is a hinge joint. Therefore, pronation and supination do not take place at the elbow, but at the radioulnar joints. Resisted dorsiflexion of the wrist causes pain in the lateral epicondyle when tennis elbow is present. Resisted flexion of the wrist causes pain in the medial epicondlye in golfer's elbow.

Injections

The elbow may be injected by a dorsal or anterolateral approach (Fig. 3–10). Using the dorsal approach, the needle is pushed forward over the tip of the olecranon with the arm flexed to 90°. The anterolateral approach requires the elbow to be flexed at 90°. The needle should be directed between the inferior

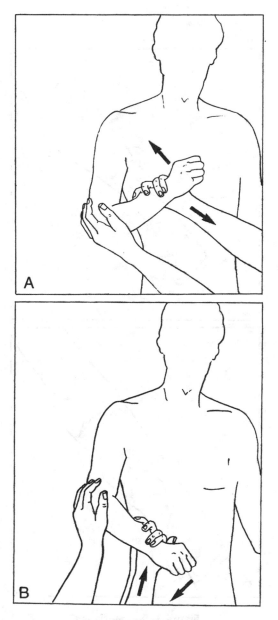

Figure 3–9.   Muscle tests for the elbow. (A) Flexion; (B) Extension; (C) Supination; and (D) Pronation.

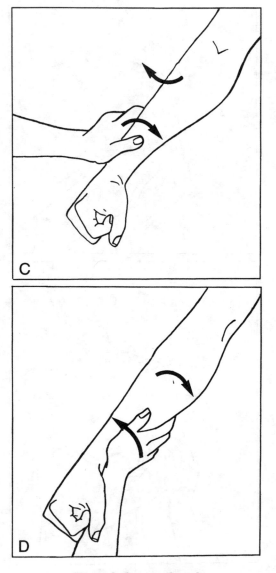

Figure 3–9—*continued.*

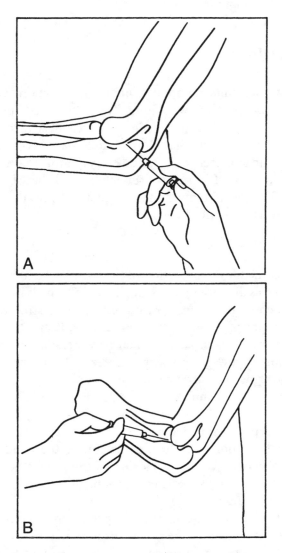

Figure 3–10. Intraarticular injection of the elbow. Injection of the elbow joint may be done by (A) a dorsal or (B) a lateral approach.

surface of the lateral epicondyle and the head of the radius. The needle should not be inserted between the two bones, but just far enough to penetrate the capsule. In an inflamed joint, the anterolateral approach is easily done and is the approach that we prefer.

HAND AND WRIST

Disease of the wrist results in impairment of hand function, which can be restored with arthrodesis of the wrist joint. The hand is perhaps the most "human" part of the body. Without the pinching action between the thumb and the index finger it is unlikely that Homo would have become Sapiens. This action enables man to inspect objects and carry out fine movements. Soldiers wishing to be invalided from active service learned that shooting off their thumb or forefinger would injure them to such a degree that they would be dismissed from active service.

Inspection

Much can be learned from inspection of the hands, which the late Dr. Philip Hench described as the patient's "visiting card." The nails may show changes of psoriasis ("pepper-pot" pitting, onycholysis, nail dystrophy); finger clubbing; nail dystrophy in the nail-patella syndrome; Beau's transverse lines or ridges due to repeated disturbances of nail growth; koilonychia in severe iron deficiency anemia; and vasculitis with splinter and nail fold hemorrhages in rheumatoid arthritis and other connective tissue diseases. Raynaud's phenomenon is frequent in connective tissue disease, especially progressive systemic sclerosis.

The position in which patients hold their hands may indicate a diagnosis. The normal position is slight dorsiflexion of the hands, with slightly flexed fingers and concave palms. Flexion of the fingers occurs in diabetic cheirarthropathy, so that the patient cannot bring the hands and fingers together as in prayer (prayer sign).

There are three palmar arches, including the carpal, metacarpal, and longitudinal. The flexor tendons lie in the grooves between the heads of the metacarpals. The arches are sustained by the intrinsic muscles. When these muscles are weak, the palmar aspect of the hand becomes flat.

The size and shape of the hand are characteristic in acromegaly, in which the fingers are spade-like; Marfan's syndrome, in which the fingers are long and thin (spider fingers), and the mucopolysaccharidoses, in which the palms are thick and the fingers short (bear palms). Heberden's nodes [first de-

scribed by the English physician William Heberden (1710–1801) as digitorum nodi] are typical and usually easy to diagnose. Osteoarthritis affecting
the proximal interphalangeal joints [first described by the French physician
Charles Jacques Bouchard (1837–1915)] may result in difficulty if Heberden's nodes are present. The distribution of osteoarthritis and rheumatoid
arthritis in the hands and wrists is important (Fig. 3–11). Subcutaneous cystic swellings containing gelatinous mucoid material may form over Heberden's nodes. These develop from the joint, like miniature Baker's cysts, but
do not contain protein. They can become painful and inflamed and have
been mistaken for acute gouty arthritis.

Fibrofatty subcutaneous pads may lie over the proximal interphalangeal
joints and be mistaken for arthritis. They are known as Garrod's fat pads, after Sir Alfred Baring Garrod (1819–1907). Juvenile rheumatoid arthritis and
psoriatic arthritis frequently involve the distal interphalangeal joints. The involvement of the carpometacarpophalangeal joint of the thumb in osteoarthritis is still a mystery.

Inspection of the hands may reveal characteristic appearances, which
give a spot (Blick) diagnosis. For example, ulnar deviation at the metacarpophalangeal joints and swan neck and buttonhole (boutonnière) deformities (Figs. 3–12 and 3–13) of the fingers indicate rheumatoid arthritis. A

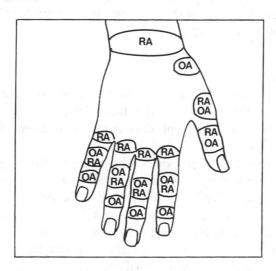

Figure 3–11. Distribution of osteoarthritis and rheumatoid arthritis in the
hands.

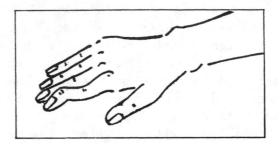

Figure 3–12. Typical "swan neck" deformity in rheumatoid arthritis. Note how there is hyperextension of the proximal interphalangeal joint and flexion of the distal interphalangeal joint.

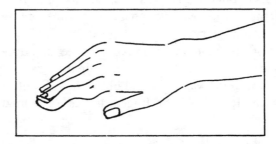

Figure 3–13. Typical "buttonhole" deformity in rheumatoid arthritis. Note the flexion of the proximal interphalangeal joint and hyperextension of the distal interphalangeal joint.

sausage-shaped finger indicates psoriatic arthritis or Reiter's syndrome. Swan neck deformity is due to contracture of the intrinsic muscles and the buttonhole deformity to the rupture of the weakened central tendinous slip of the extensor hood.

A Z-shaped, or hitchiker's, deformity of the thumb (Fig. 3–14) is common in rheumatoid arthritis, although it may also be seen in other inflammatory joint diseases. Tendon rupture in rheumatoid arthritis, especially of the lateral extensor tendons of the hand, gives rise to a clergyman's hand (Fig. 3–15). Swelling of a single phalanx used to be seen in tuberculosis and syphilitic dactylitis, but is rare today. Sickle-cell dactylitis, however, may occur and cause painful swelling of a phalanx.

Muscle wasting in the hand may be generalized or localized. When localized, it may affect the thenar and hypothenar muscles. The ulnar nerve

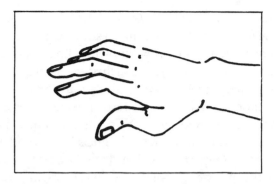

Figure 3–14. Typical Z-shaped or hitchhiker's deformity of the thumb in rheumatoid arthritis.

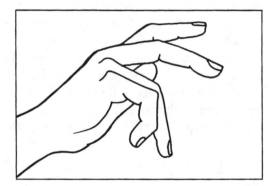

Figure 3–15. Rupture of fourth and fifth tendons of extensor digitorum communis.

supplies the hypothenar muscles, and the median nerve supplies the thenar eminence muscles. The first interosseous and adductor pollicis brevis are supplied by a branch of the ulnar nerve.

Anhydrosis may occur with damage to a nerve. Hyperhydrosis is usually due to nervousness and a hot environment, but may occur when a nerve is irritated or when the patient has thyrotoxicosis.

Palpation

The clinician should be acquainted with the anatomical landmarks of the hand and wrist and should be able to palpate the radial and ulnar styloids, scaphoid tubercle, and pisiform and hook of the hamate. The anatomical relationships of the fovea radialis (anatomical snuff box) and extensor tendons on the dorsum of the hand should be known. Palpation of these tendons is important in patients with rheumatoid arthritis, especially because synovitis may result in rupture. The extensor tendons may contain rheumatic fever nodules and xanthomata. These can be easily missed if they are small and the extensor tendons are not examined with the hand clenched (Fig. 3–16). Inflammation of the tendon sheaths of the pollicis longus and extensor pollicis brevis may result from repeated friction. If the process, known as de Quervain's disease, is chronic, stenosis of the tendon sheath may result from thickening. A local injection of corticosteroid into the tendon sheath is the treatment of choice. Diagnosis can be made by reproducing the pain by forcible ulnar deviation of the wrist.

The flexor tendons should also be palpated, because synovitis may result in swelling with resulting difficulty in moving the fingers and, rarely, tendon rupture. A trigger finger may result from tendon nodules, which can be palpated at the metacarpal heads. It is these nodules that are trapped in the fibrous tendon sheath, preventing extension of the finger (Fig. 3–17). A mallet finger is due to the rupture of the long extensor of the finger at its insertion into the base of the distal phalanx. Although it may occur in rheumatoid arthritis, it is more commonly the result of trauma.

Compression of the carpal tunnel is perhaps one of the most common

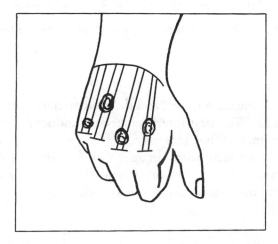

Figure 3–16. Xanthomata are best seen in the dorsal tendons when the patient makes a fist.

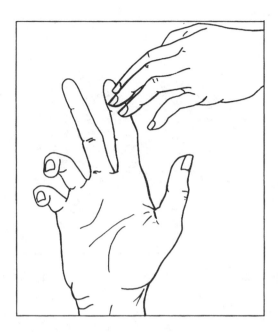

Figure 3–17. Trigger finger. During extension or flexion of the finger, an audible and palpable snapping may occur. This is caused by a nodule in a flexor tendon catching on the narrow annular sheath or pulley opposite the metacarpal head, as shown. A trigger thumb may also occur.

conditions seen by the family doctor (7). The anatomical relationships are illustrated in Figure 3–18. Compression of the tunnel (or, more correctly, the canal) may be caused by a variety of conditions (Table 3–1). The areas of the hand that the median and ulnar nerves supply are shown in Figure 3–19. In healthy subjects, there is considerable variation in the areas innervated by these nerves. Figure 3–19 also shows the area innervated by the radial nerve. As Cailliet points out, it is important to remember that these are the "usual" motor and sensory distributions of the nerves (8). An all-median or all-ulnar nerve hand may occur. The radial nerve may have a small sensory distribution on the dorsum of the hand and, occasionally, no distribution. Accurate diagnosis can only be made by an electromyography (EMG) (8). The median nerve, when compressed (Fig. 3–20), results in pain, numbness, and paresthesia in the area innervated by the median nerve (Tinel's sign). Compression of the hands by the patient, as shown in Figure 1–5, causes the same symptoms (Phalen's sign).

The distal radioulnar joint may give rise to pain on the volar aspect of the

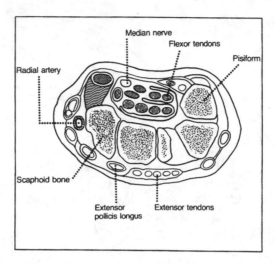

Figure 3–18.   Anatomical relationships of carpal tunnel.

Table 3–1. Causes of Median Nerve Compression in the Carpal Tunnel

1. Idiopathic
2. Pregnancy
3. Acute trauma [e.g., dislocation of the lunate (all carpal bones, with the exception of the lunate, dislocate in a dorsal direction)] and chronic trauma (e.g., following a Colles fracture)
4. Rheumatoid arthritis and other inflammatory arthritides
5. Localized swellings (e.g., gouty tophus and amyloid deposits) common in the light-chain form of multiple myeloma and β-2 microglobulin amyloid in chronic dialysis patients
6. Diabetes mellitus
7. Hypothyroidism
8. Acromegaly

wrist in the midline. When the distal or proximal radioulnar joint is involved, difficulty and pain are encountered when the hand is pronated, such as when counting money (Rooney's sign). The greatest amount of synovium in the wrist is around the carpoulnar joint. Therefore, pain and tenderness are most marked around the ulnar head. Consequently, erosion of the fibrocartilage between the ulnar and the triquetrum occurs and results in upward dislocation of the ulnar head. It is often possible to push the ulnar head up and down (the piano sign).

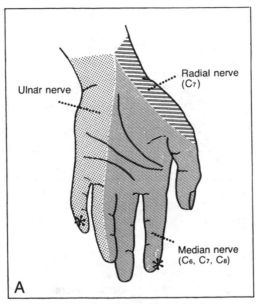

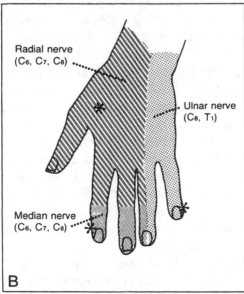

Figure 3–19. Distribution of median ulnar and radial nerve supply to (A) palmar and (B) dorsum aspects of the hand, including areas marked by stars of almost autonomous innervation.

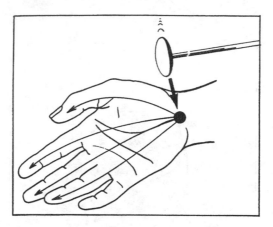

Figure 3–20. Tinel's sign. Pain is elicited over the distribution of the median nerve by tapping over the volar carpal ligament.

Dupuytren's contracture of the palmar fascia produces a typical lesion and flexion contracture of the fingers, especially the fourth and fifth fingers. The contracture can occur in healthy people, but is more frequent in patients with alcoholic hepatic cirrhosis and diabetes mellitus.

The carpometacarpal joint of the thumb is commonly involved in osteoarthritis. The patient frequently cannot abduct the thumb, which is held close to the index finger. This results in a square hand, sometimes referred to as the shelf sign. The joint is tender and crepitus can be elicited.

The metacarpophalangeal joints are frequently swollen and tender when inflammatory polyarthritis is present. In early rheumatoid arthritis, the second and third joints are commonly involved. The involvement of these joints is also common in hemachromatosis. In rheumatoid arthritis, a double swelling may be seen on the extensor surface of the metacarpophalangeal joints, which may be bisected by the extensor tendon. Pigmentation may be apparent over inflamed metacarpophalangeal and proximal interphalangeal joints in people with dark skin.

Swelling of the proximal interphalangeal joints may give rise to spindling of the finger. This is often considered pathognomonic of rheumatoid arthritis, but can also be seen in other inflammatory joint diseases and osteoarthritis (Bouchard's nodes). Buttonhole and swan neck deformities are also common in rheumatoid arthritis, but are not pathognomonic for the disease because they occur in other forms of inflammatory polyarthritis.

The skin of the hands may show the typical appearance of progressive systemic sclerosis. The early stages of this disease involve swelling, diffuse erythema, and, occasionally, pitting edema may occur. Later, the skin becomes tight and shiny and fixed to the underlying bone. The patient may experience Raynaud's phenomenon. Ulceration and dry gangrene may result from digital artery ischemia. Subcutaneous calcification and osteolysis of the terminal phalanges also occur. The skin of the hands and the rest of the body may be deeply pigmented. Vitiligo may also occur. Crepitus may be audible with a stethoscope over involved tendon sheaths due to fibrin deposits.

In addition to these characteristic features of progressive systemic sclerosis, the skin lesion in dermatomyositis may also be a typical manifestation. The lesion consists of linear erythema along the dorsum of the fingers (Groton's sign). Vasculitis nodules may be found in polyarteritis nodosa and other connective tissue diseases, including rheumatoid arthritis and systemic lupus erythematosus. These vasculitis nodules may ulcerate and take many months to heal. Vasculitic lesions may also be found around the finger nails. Periungual erythema is commonly seen in active connective tissue diseases. Osler's nodes [first described by Sir William Osler (1859–1919)] are a feature of subacute bacterial endocarditis. Palmar erythema may be seen in severe rheumatoid arthritis and systemic lupus erythematosus.

Rheumatoid nodules may be present on the fingers and, like xanthomata, be intracutaneous. Reticulohistiocytosis may also give rise to nodules over the finger joints and may be mistaken for rheumatoid arthritis when articular erosions are present. Nodules may also be present over the ears. In chronic gouty arthritis, tophi are common on the pinna of the ears, but may occur at many different sites where they can be mistaken for rheumatoid nodules. When tophi ulcerate through the skin, they exude a white chalky material. Under the microscope, this material can be seen to consist of masses of sodium urate crystals. Sir Arthur Conan Doyle (1859–1930) recorded in one of his Sherlock Holmes' stories how Lord Mount James, "one of the richest men in England . . . could chalk his billiard cue with his knuckles!"

Movements

Tests for movement in the hands and wrists are simpler than those for the shoulder and elbow. The movements of the fingers are finer and the anatomical reasons for inability to perform simple movements are more difficult to explain. In addition to flexion and extension of the finger joints and abduction of the wrist, it is also important to test the pinching strength of the thumb with each of the fingers. This can be done by drawing a sheet of paper through the opposed thumb and finger.

The palmar interossei adduct can be tested by drawing a paper through

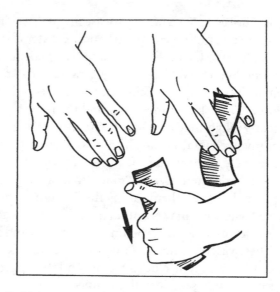

Figure 3–21.   A simple method of testing finger adduction.

the fingers opposed to each other (Fig. 3–21). Circular and lateral movement normally occurs at the metacarpophalangeal joints when they are in extension, not flexion. Subluxation of the metacarpophalangeal joints in rheumatoid arthritis can be easily recognized on clinical examination.

Movements in the hands have been divided into two categories; precision grip and power grip (9). Precision grip tests the various types of pinch, whereas power grip tests muscle strength while grasping.

The most common form of ganglion occurs over the dorsum of the wrist. It contains mucinous material and, although usually soft and fluctuant, may be sufficiently hard to be mistaken for bone. It usually disappears with repeated needling or injection of corticosteroid. If this treatment fails, surgery may be required. The second form of ganglion was formerly seen with tuberculous infection, but is now more commonly seen with rheumatoid arthritis. The compound palmar ganglion presents with cross-fluctuant swellings on either side of the flexor retinaculum of the wrist. In tuberculosis and rheumatoid arthritis, the ganglion may contain rice bodies.

Extreme mobility of the wrist and hand joints may occur in Marfan's syndrome, Ehlers-Danlos syndrome, and other rare heritable disorders of connective tissue. If neuropathic arthropathy is present, abnormal and painless movement may occur in the wrist or joints in the upper limb. The most common cause in the upper limb joints is syringomyelia. In the "main en

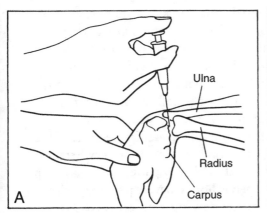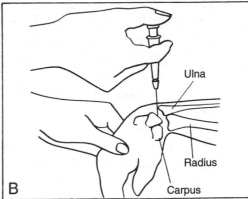

Figure 3–22. Technique of intraarticular injection of the wrist. (A) Dorsoradial approach; and (B) Dorsoulnar approach. The wrist can be injected on the ulnar or radial side using a dorsal approach with the wrist in the flexed position.

lorgnette" (e.g., arthritis mutilans in rheumatoid arthritis, psoriatic arthritis) the joints are totally destroyed, resulting in marked hypermobility. In osteoarthritis, the distal interphalangeal joints of the fingers may also demonstrate instability. In our experience, this occurs especially in the erosive form of osteoarthritis.

Injection

Injection of the wrist joint can be performed by a dorsoradial or dorsoulnar approach (Fig. 3–22). In the dorsoradial approach, the needle is inserted vertically, just distally to the ulnar side of the radial styloid. In the dorsoulnar approach, the needle is inserted immediately on the radial side of the styloid process of the ulna. The needle should not be inserted through tendons. Injection of other joints, tendon sheaths, and carpal tunnel are well described in the book by Docherty et al. (10).

HIP JOINT

The current focus is on the hip joint due to the spectacular results of hip joint arthroplasty. Even mild disease of the hip joint causes severe pain. Confusion may arise as to the origin of the pain because hip pain can simulate pain arising from sciatic nerve irritation and can be referred to the knee joint. Patients often experience knee joint pain because the hip and knee joints are both innervated by branches of the obturator and femoral nerves. Osteoarthritis is the most common disease affecting the hip joint and may cause severe disability. Although the hip joint is relatively spared in rheumatoid arthritis, when it is involved it can severely cripple the patient. In the past, both physician and patient were demoralized if the cause of disability was severe hip disease. Today, effective therapy can be offered by the orthopaedic surgeon.

A number of diseases can affect the hip joint. Avascular necrosis of the femoral head, most often seen in patients with systemic lupus erythematosus, can affect the hip joint. A middle-aged or elderly patient may have hip pain as a result of Paget's disease of the pelvic bones. Severe pain may arise in Paget's disease as a result of fracture or the development of osteosarcoma. Stress fractures of the femoral neck can cause hip pain in young patients unaccustomed to active exercise and in elderly females as a result of osteoporosis. Osteomalacia may also cause fractures involving the pelvic bones and femur (Looser's zones) and results in a waddling gait. A peculiar type of acute osteoporosis can occur in adult males, which is probably related to regional migratory osteoporosis. This acute osteoporosis presents with severe pain and stiffness in the hip, which is thought to be due to osteoarthritis.

Perhaps the most common cause of hip pain in children is "irritable" hip, especially in young boys, which is probably due to a traumatic synovitis. The condition is benign and self-limiting and usually clears up in a few days. Therefore, it is often referred to as "observation" hip. The only finding on radiograph is a dense shadow lateral to the femoral head, indicating bulging of the capsule due to effusion. It is difficult to differentiate this benign condition from other causes of hip disease, especially septic arthritis. Fever, leucocytosis, and an elevated erythrocyte sedimentation rate help to differentiate septic arthritis from "irritable" hip. If septic arthritis is considered, joint aspiration by needle or surgical drainage should not be delayed, because the pressure of the septic effusion may compromise the blood supply to the femoral head.

Hip involvement is relatively common in juvenile rheumatoid arthritis. Although it often occurs early in the disease, it is rarely monoarticular. Legg-Calvé-Perthes disease, also known as coxa plana, occurs primarily in boys between the ages of four and eight years. It is frequently bilateral. A radionuclide bone scan is useful in early diagnosis, showing a decreased uptake due to ischemia. Early recognition of the disease and institution of treatment can prevent epiphyseal damage. Congenital dislocation of the hip is usually bilateral and can be tested for in infancy by several maneuvers. Subluxation of the femoral head upward and outward can be easily recognized by interruption of Shenton's line on radiograph. This is a continuous line from the neck of the femur to the pubic ramus. Acetabular dysplasia can be measured radiologically by measurements of various angles, which are well described in radiologic texts. In adults, a dislocated hip or coxa vara can be diagnosed if the femoral head lies above Nelaton's line (from the anterior superior iliac spine to the ischial tuberosity).

Inspection

Hip joint disease can often be diagnosed in the consulting room. The patient attempts to carry his weight on the uninvolved limb and uses a cane for support. The cane should always be held in the hand opposite the affected side, which can reduce the load on the affected hip by 40%. The affected limb is dragged along and supports the body weight for the shortest possible time.

The patient is often asked to sit and rise from a chair in a characteristic manner. Putting on shoes and stockings is often performed with the patient sitting and bending the knee backward. Although a waddling gait is typical of bilateral congenital hip displacement, it may occur in a number of diseases. Inspection of the hip joints includes recording of any superficial abnormalities, such as discoloration of the skin, swelling, sinuses, and herpes zoster. It is important to note whether the anterior superior iliac spines, greater trochanters, and gluteal clefts are level. The limb lengths and the compared level of the knees should also be noted. When examining the hip joints, it is essential to examine the lumbar spine. Scoliosis of the lumbar spine may cause apparent shortening of a limb. The position of the feet is also important because severe hip joint disease and fractures of the femoral neck may cause eversion of the foot.

It is important to ascertain whether or not leg shortening is apparent or true. This is done by measuring from the anterior superior iliac spine to the medial or lateral malleolus. This should demonstrate whether or not there is true leg shortening. In practice, this is difficult because patients with hip disease often have flexion contracture. An alternative method for differentiating true and apparent leg shortening is to raise the foot until the anterior superior iliac spines lie in the horizontal plane. This method may also be ineffective if the patient has a flexion contracture of the hip joint.

If leg length shortening is present (differences greater than 1.0–1.5 cm being considered abnormal) it is important to determine the reason. Measurements should be made from the iliac crest to the greater trochanter of the femur on both sides. If differences are found, this will indicate whether the shortening is due to hip joint disease (e.g., coxa valga or vara). To determine whether there is femoral shaft shortening, measurement should be made from the greater trochanter of the femur to the lateral aspect of the knee joint line. Tibial shaft shortening may be apparent with the patient prone and the knees flexed at 90°. In this scenario, the knee on the shortened side will be lower. Measurement can confirm tibial shaft shortening by measurement from the medial side of the knee joint line to the medial malleolus.

Palpation

The clinician should be acquainted with the four anatomical landmarks around the hip. These include the anterior superior iliac spine, symphysis pubis, greater trochanter of femur, and the ischial tuberosity. The upper border of the symphysis pubis and the greater trochanter are normally level. Palpation in the iliac fossa should include palpation of the femoral artery, inguinal lymph nodes, and femoral and inguinal herniae. The iliopectineal bursa may be mistaken for a femoral hernia. The iliopectineal (or ilipsoas) bursa often communicates with the hip joint. It is not the first femoral arteriogram that has produced a hip arthrogram. Effusion in the hip joint causes an anterior bulging in the inguinal fossa. In thin subjects, the iliopsoas muscles may be palpated through the abdomen. Medial to the femoral artery lies the femoral vein, and the femoral nerve lies laterally. The former may be tender in deep venous thrombosis of the leg, and the latter in disc lesions affecting L3 and L4 nerve roots. The femoral pulse may be diminished in coarctation and severe atherosclerotic disease of the aorta, both of which may give rise to intermittent claudication and gluteal pain.

The adductor muscles are easily palpated medial to the femoral vein, and may be tender and spastic. Adductor tendinitis is common in athletes, especially horse riders ("rider's strain"). Stretching of the adductor longus tendon by abducting the leg causes intense pain. In hip joint disease, the obturator nerve may be tender on palpation because one of its branches supplies the joint. Because the obturator nerve innervates the adductor muscles, painful reflex spasm may result. Lateral to the femoral artery and nerve, and medial to the sartorius muscle, the anterior aspect of the hip joint may be palpated. This is also the most appropriate site for intraarticular injection of the hip joint. When severe hip joint disease with flexion contraction is present and attempts are made to extend the hip, pressure effects may occur on the ilioinguinal nerve. This causes pain in the inguinal fossa and on the lateral cutaneous nerve of the thigh, which results in pain down the lateral aspect of the thigh. The lateral cutaneous nerve of the thigh, derived mainly from L2 and L3 nerve roots, supplies the skin over the anterolateral aspect of the thigh. It is sometimes compressed in a tunnel in the inguinal ligament close to the attachment to the anterior superior iliac spine. The patient experiences burning pain, paresthesias, and numbness over the nerve distribution, known as meralgia paresthetica. Injections of local anesthetic and corticosteroid one finger breadth medial and below the anterior superior iliac spine frequently alleviate the symptoms. In ankylosing spondylitis and psoriatic arthritis, periostitis around the symphysis pubis and inferior pubic ramus may be tender on palpation. Likewise, periostitis may cause tenderness on palpation over the greater and lesser trochanters.

Pain may also arise in the symphysis pubis due to instability following operations on the sacroiliac joint, childbirth, and trauma. Instability is a common sporting injury, especially in soccer players and runners. Diagnosis is confirmed by radiograph with the patient standing on one leg. According to the definition, instability exists when the heights of the two sides of the pubis exceed 2 mm.

In severe osteoarthritis of the hip, the external rotator and abductor muscles may be taut and cause tenderness when their insertions into the greater trochanter are palpated. A trochanteric bursitis may cause tenderness above and around the trochanter, but swelling is seldom palpated. The fascia lata may be tender on palpation when there is irritation of the lateral cutaneous nerve of the thigh, which may occur in severe flexion of the hip joint.

Palpation of the buttock is best performed with the patient lying on her side with the hip and knee joint flexed (Sim's position). This causes the gluteus maximus to move upward, allowing the ischial tuberosity to be easily palpated. The ischial tuberosity is frequently tender in ankylosing spondylitis, psoriatic arthritis, and Reiter's disease, due to periostitis. It may also be tender when the hamstrings are taut. The tendinous origin of the hamstring muscles from the ischial tuberosity frequently becomes tender due to inflammation from overuse. It is common in distance runners. Separation of the tendon from its insertion may occur in sprinters ("sprinter's fracture") and is readily evident on radiograph. Tenderness over the ischial tuberosity can also occur in chronic inflammation of the ischiogluteal bursa ("weaver's bottom"). The sciatic nerve can be palpated midway between the greater trochanter and the ischial tuberosity. The nerve may be tender in disc disease or any other condition exposing it to pressure.

The gluteal muscles are frequently wasted in severe hip disease. In addition, they may be taut and tender, which is aggravated on movement. Hypertrophy and pseudohypertrophy of the gluteal muscles may occur in myopathy (e.g., progressive muscular dystrophy). The sacroiliac joint can only be palpated posteriorly and may be tender when sacroiliitis is present. The iliac bone should be palpated to ensure that tenderness over the sacroiliac joints is not due to iliac bone disease, such as Paget's disease of bone. In elderly females, there may be subcutaneous tenderness overlying the buttock and lateral thigh due to panniculitis. When the overlying skin is compressed between the thumb and index finger a typical orange peel appearance results and the lobulated subcutaneous fatty tissue can be palpated.

Movements

Movements of the hips should be performed in three ways; active, passive, and resisted isometric movements. The movements to be tested should include flexion (120°), extension (15°), abduction (50°), adduction (30°), lateral rotation (60°), and medial rotation (40°). The figures in parenthesis indicate the upper limit of the normal range during active movements. A greater range can be expected with passive movements (e.g., extension of the hip to 30°).

A quick screening test for hip joint movement is the Patrick's test. It is also known by the acronym, FABER, which stands for Flexion, Abduction, and External Rotation. With the patient supine and the hip and knee flexed, the knee is moved laterally toward the floor. This movement abducts and laterally rotates the hip. The test does not provide information regarding specific hip movements.

The Trendelenburg test is useful in eliciting weakness of the gluteus medius. This is performed by observing the dimples overlying the posterior superior iliac spines. When the patient raises his leg on the unaffected side, the dimple on this side remains horizontal or is depressed. Normally, the dimple should rise (Fig. 3–23). The Trendelenburg test may be positive in a number of conditions, including congenital dislocation of the hip, coxa vara, paralyzed hip abductor muscles, and painful conditions about the hip. An alternative procedure for testing weakness of the gluteus medius is to ask the patient to raise her leg upward while lying on her side. This movement is performed almost entirely by the gluteus medius. Painful movement is due to gluteus medius tendinitis. Painless weakness indicates an S1 nerve root lesion or rupture of the gluteus medius tendon.

The iliopsoas is the primary flexor of the hip joint, with some contribution from the rectus femoris. The flexion test for the iliopsoas is best performed with the patient sitting at the edge of the couch with the legs dependent. The patient is then asked to raise the knee against resistance. When the patient cannot raise the leg with an extended knee, a fracture of the femoral neck should always be considered, especially in an elderly person. Extension power of the gluteus maximum is best tested with the patient lying prone (Fig. 3–24). The patient is asked to raise the leg while the examiner exerts pressure on the hamstrings. The examiner can test the power of the gluteus maximus with the other hand. The abductors (gluteus medius and, to a lesser extent, gluteus minimus) and adductors (adductor longus and, to a lesser ex-

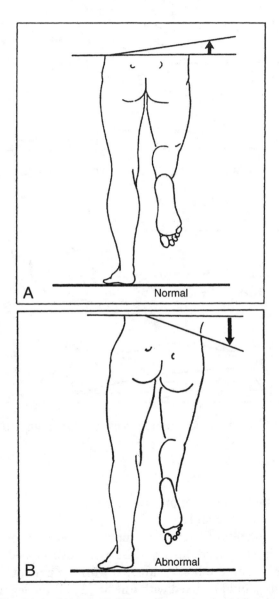

Figure 3–23. Trendelenburg test. When the gluteus medius is weak, the dimple over the posterior superior iliac spine remains horizontal or is depressed (positive test).

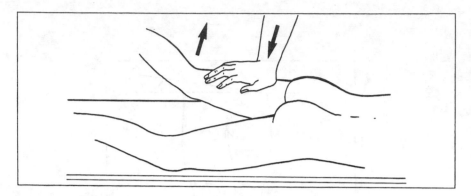

Figure 3–24. Testing gluteus maximus muscle power.

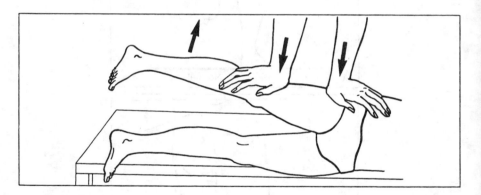

Figure 3–25. Muscle tests for abduction (gluteus medius and, to a lesser extent, gluteus minimus).

tent, adductor brevis, adductor magnus, pectineus, and gracilis) can be tested as shown in Figures 3-25 and 3-26.

Passive movements of the hip are important because contractures of the hip joint can develop insidiously and be easily overlooked. Passive movements of the hip joint are best conducted with the patient reclining on his back. The pelvis should be fixed with one hand and the leg moved with the other. Abduction and adduction require no special comment. Flexion is tested by flexing the hip and knee joints (Fig. 3–27). A hidden flexion contracture of the opposite hip will become apparent when the knee joint on the opposite side flexes (Thomas's sign). External and internal rotation may be examined with the knees extended or flexed (Fig. 3–28). Unlike other movements, extension of the hip joint is only tested when the patient is lying prone (Fig. 3–29). The examiner uses one hand to fix the pelvis and the other to extend the leg.

In infants, congenital dislocation of the hip can be tested by four methods; Ortolani's sign, Barlow's test, Galeazzi's sign (or Allis' test), and the Telescoping sign. Ortolani's sign is a click on abduction and lateral rotation, and is valid only for the first few weeks after birth. Barlow's test is a modification of the Ortolani maneuver and is valid up to 6 months after birth. Galeazzi's

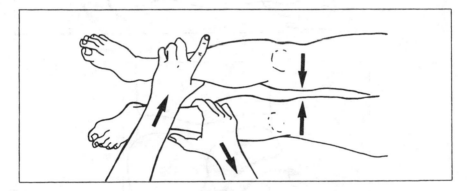

Figure 3–26. Test for adductor muscle strength (adductor longus and, to a lesser extent, adductor brevis, adductor magnus, pectineus, and gracilis).

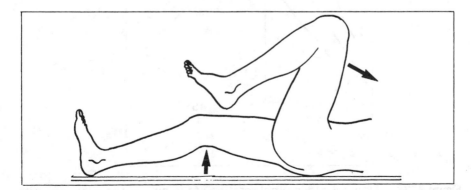

Figure 3–27. Thomas test for flexion contracture of the hip. When the hip is fully flexed, the lumbar spine is flattened and the pelvis stabilized. If flexion contracture of the opposite hip is present, flexion of the opposite knee will result.

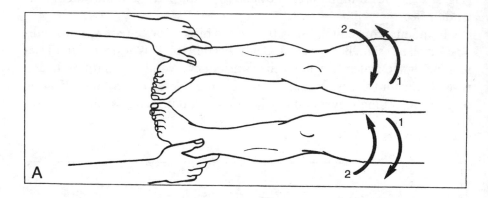

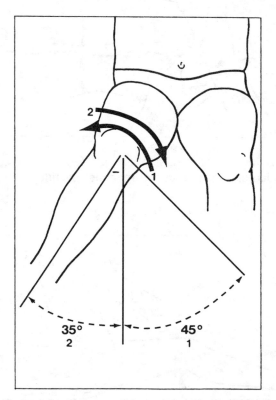

Figure 3–28.   External and internal rotation of the hip may be examined with the knee joint (A) extended or (B) flexed. External rotation is normally 45° (1) while internal rotation is 35° (2).

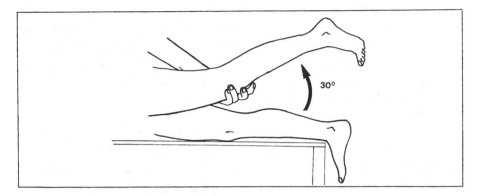

Figure 3–29. Extension of the hip. The normal range is from 0–30°.

sign, or Allis' test, is useful in detecting unilateral hip dislocation and is valid from 3 to 18 months of age. Excessive movement of the hip, called telescoping or pistoning, is evident in a child with a dislocated hip. These various tests, and those used for adults, are superbly described by Magee (6).

Injection

Injection of the hip joint may be made by an anterior or lateral approach. We prefer the anterior approach, with insertion of the needle two fingers width lateral to the femoral artery and just below the inguinal ligament. In the lateral approach, the injection is given directly above the tip of the greater trochanter. Ideally, injection of the hip joint should be performed under fluoroscopy.

KNEE JOINT

The knee joint is not only the largest joint in the body, but also the one most subject to trauma. Due to its large amount of synovium, it is also commonly affected in inflammatory joint disease. It is also one of the most commonly affected joints in osteoarthritis.

Inspection

Movements of the knee should be observed while walking and sitting down and rising from a chair. Abnormal ligamentary laxity may be instantly apparent. Genu valgum, genu varus, genu recurvatum, and flexion deformity are easily recognized. Genu valgum (knock-knees) can be diagnosed when the distance between the ankles exceeds 10 cm. If the ankles touch, but the knees do not, the patient has genu varum (bowlegs). The tibia may be curved in Paget's disease, rickets, and syphilis (sabre tibia). Erythema over the knee joint may be marked in septic arthritis and gouty arthritis. Rarely, erythema is found in other joint diseases. Psoriatic lesions may be seen over the patella and tibial tuberosity.

Swelling of the knee itself must be differentiated from swelling around the joint, such as a Baker's cyst, prepatellar bursitis, prominence of the tibial tuberosity in osteochondritis (Osgood-Schlatter's disease), hyperlipoproteinemia, a meniscal cyst, an exostosis, a Pelligrini-Stieda lesion, or a tumor. A meniscal cyst usually is associated with the lateral meniscus and contains thick, gelatinous material. It is usually tender and is best seen when the patient is sitting with the knee flexed at 45°. These cysts are usually associated with traumatic tears in the meniscus. A Pellegrini-Stieda lesion produces a tender ligament to the femoral condyle. It is due to hematoma as a result of injury. A radiograph will show a line calcification running parallel to the femoral condyle. Displacement of the patella may be seen in genu valgum and in the rare condition of congenital dislocation of the patella. Atrophy of the quadriceps muscle is important in disorders of the knee joint and can be the cause of joint pain and swelling. The vastus medialis muscle is responsible for the "screw-home" movement in the last 15° of knee extension of the femur and is the muscle of the quadriceps group that demonstrates most atrophy. Formerly, atrophy of the calf muscles was most commonly due to anterior poliomyelitis. Today, in the absence of this disease, other conditions should be considered [e.g., progressive neuropathic peroneal muscular atrophy (Charcot-Marie-Tooth syndrome)].

Palpation

Palpation may reveal the presence of a small patella or even its absence in the nail-patella syndrome. A bipartite patella may also be palpated. The patella may be abnormally high (patella alta) or lower than normal (patella baja). When a high patella is viewed from the side, two humps can be seen ("camel sign"). One of the humps is the high-riding patella and the other the intrapatellar fat pad. The joint space is best palpated on the medial aspect of the knee. Osteophytes may be palpated and hypertrophied synovium may protrude between the joint margins. Chronic synovial hypertrophy is best palpated along the superior margins of the suprapatellar pouch and gives a characteristic rubbery sensation as it is rolled under the fingers. The intraarticular fat pad of Hoffa may be palpated on both sides of the patellar ligament.

In a young patient with chondromalacia patellae, crepitus may be felt over the patella on movement of the joint. Although crepitus is usually fine, in an elderly patient with severe osteoarthritis it is coarse. An audible and palpable click may be present in a number of conditions affecting the knee, including a synovial plica. Fluid in the knee joint may be detected by a patellar tap (Fig. 1–2) or by milking the fluid from the lateral to the medial side of the joint. A ripple appears just medial to the patella (Fig. 1–2). Fluctuation may be apparent in a prepatellar bursitis. A Baker's cyst most often occurs in the popliteal fossa, but may appear on either side of the joint or anteriorly. The patient should always be lying prone when examining the popliteal fossa.

Since confusion often exists as to what exactly a Baker's cyst is, a definition of the term will be useful. William Morant Baker (1839–1896), a surgeon at Saint Bartholomew's Hospital in London, described cysts in the popliteal fossa in disease of the knee joint (mostly tuberculosis). It is now known that these cysts are due to herniation of the synovial membrane through the capsule of the joint and distension of the semimembranosus bursa, which communicates with the knee joint in 40% of normal subjects. There are several important bursae around the knee.

Semimembranosus bursae are located in three areas. They can be found between the tibial collateral ligament and tendon of insertion of semimembranosus; between the semimembranosus tendon and medial tibial condyle; and between the medial head of gastrocnemius and the overlying semimembranosus muscle. Bursa anserina can be found deep between the tendons of the sartorius, gracilis, tibia, and the tibia collateral ligament. Another type of bursa are those situated between the inner head of gastrocnemius and medial femoral condyle. Rarely, bursa can be found between the tendons of semimembranosus and semitendinosus. Deep infrapatellar bursa can be found between the tibial tuberosity and ligamentum patellae. It is separated from the knee joint by the infrapatellar fat pad. When inflamed, tender swellings appear on either side of the insertion of the patellar ligament. Superficial bursae include the superficial prepatellar bursa, which often becomes inflamed (housemaid's knee) and the superficial infrapatellar bursa, which can also become inflamed (clergyman's knee).

Small cysts may be easily missed if the popliteal fossa is examined with the patient lying on her back. A Baker's cyst is often tense and may occupy the entire popliteal fossa or track down into the calf and appear as a large swelling at the posterior aspect of the ankle joint. The calf swelling is often more evident in the anterior medial aspect in the upper third of the calf. Unlike a prepatellar bursa, a Baker's cyst is seldom translucent.

An excellent review of synovial cysts can be found in the paper by Gerber and Dixon (11). Rupture of a Baker's cyst results in a typical clinical picture. There is immediate pain behind the knee that radiates into the calf. On palpation, the calf becomes tender and swollen. Homans' sign is frequently positive and consists of dorsiflexion of the foot, causing calf pain in the presence of deep venous thrombosis. There may be slight fever and leucocytosis. Edema is also common around the ankle. Diagnosis can readily be confirmed by arthrography or aspiration of synovial fluid from the calf. A careful history may elucidate the fact that the patient's previously swollen knee has shrunk. Therefore, the simple and frequently forgotten exercise in history taking can obviate the need for an arthrogram or aspiration. A hemorrhagic crescent sign may rarely occur below the lateral malleolus at the ankle. Arthrograms, ultrasound, and computed tomography (CT) scans are often used to confirm the presence of a ruptured Baker's cyst. The arthrogram may not detect the cyst because the communication with the joint may have closed. Ultrasound techniques to detect patency of veins may give false results because the veins are compressed in the calf and popliteal fossa by extravasation of fluid. A venogram is the only sure way to exclude a deep venous thrombosis. In our opinion, ultrasonography is the preferred test to detect a Baker's cyst. A discussion on the differential diagnosis of a ruptured Baker's cyst is given by Schmitt, et al (12).

In addition to Baker's cysts, other structures may be palpated in the popliteal fossa (e.g., aneurysms of the popliteal artery, absence or diminished pulsation of the popliteal artery, tenderness and thickening of the tibial nerve, and varicose veins).

Between the knee and the ankle there are four compartments in the leg: anterior, medial, lateral, and posterior. These compartments are formed by interosseous membranes and deep fascia, which separate the various muscle groups. The most common syndrome associated with these compartments is that involving the medial tibial one. Exercise causes pain and swelling of the muscles, and pain and tenderness over the medial tibial border. The resulting condition is known as shin splints and has to be differentiated from a stress fracture of the tibia. This can be done with a plain radiograph or radionuclide scan. A similar syndrome can affect the anterior tibial compartment and, when acute, constitutes a medical emergency. If surgical decompression is not done rapidly, permanent damage to the deep peroneal nerve may occur, with resulting foot drop. Chronic compression syndromes also may affect the lateral, posterior, and anterior compartments.

Tears of the gastrocnemius muscle often occur in middle-aged men while jogging or playing tennis ("tennis leg"). There is usually no difficulty in making the correct diagnosis.

Ligament laxity or tenderness is easily tested by putting the ligaments on stretch. Figures 3-30–3-32 illustrate how this is done. The methods for test-

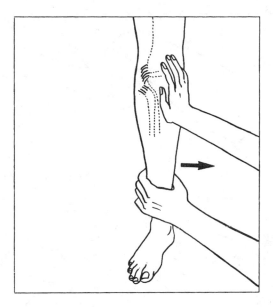

Figure 3–30.   Testing lateral ligament laxity. This is best performed with the patient lying on his back.

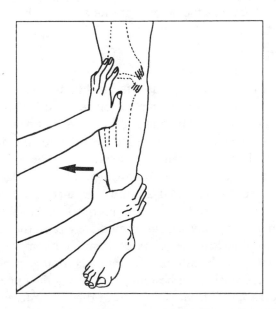

Figure 3–31.   Testing for medial ligament laxity. This is best performed with the patient lying flat on his back.

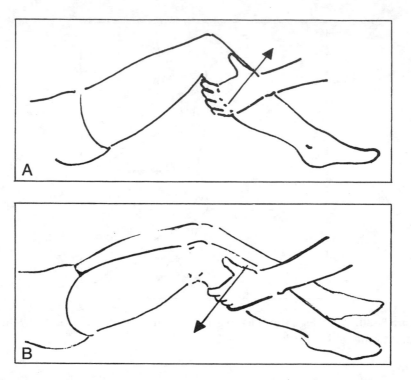

Figure 3–32. Testing for (A) anterior and (B) posterior cruciate ligament laxity. The former is known as the anterior drawer sign and the latter as the posterior drawer sign.

ing integrity of the menisci are illustrated in Figures 3-33 and 3-34. Other tests include Apley's downward compression test and McMurray's test. For full descriptions of these and the many other tests that have been described for meniscal tears, the reader is referred to the textbook by Magee (6).

Instability of the fibular head may be palpated. The peroneal nerve can be palpated for tenderness and thickening around the fibular neck. The following tendons may be palpated around the knee joint: biceps femoris, semimembranosus, semitendinosus, and prepatellar. Tibial tuberosity pain on palpation in a young boy should immediately raise the suspicion of osteochondritis (Osgood-Schlatter disease). Panniculitis may give rise to tenderness on palpation of the infrapatellar fat pads (orange peel sign).

Knee joint circumference is best measured using the upper border of the patella as a reference point. Measurement of the circumference of the thigh is useful in determining quadriceps atrophy and should be performed 10–15 cm above the upper border of the patella.

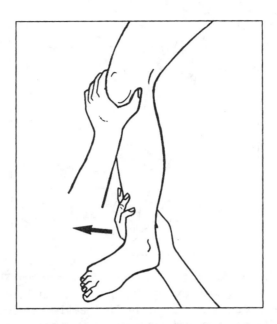

Figure 3–33. Testing for anterior lesions of the menisci. The thumb is pressed firmly into the joint line at the medial aspect of the patellar ligament. The knee is then extended. A click accompanied by pain indicates an anterior meniscus lesion. This is best performed with the patient lying flat on his back (McMurray test).

Movements

Knee movements include extension (0–15°) and flexion (0–135°). In slight flexion of the knee, a slight degree of in-and-out rotation also takes place.

Injection

The knee joint is the most frequently injected joint in the body. The most commonly used sites are on the medial and lateral sides of the knee, just below the middle of the medial or lateral edge of the patella (Fig. 3–35). Pushing down on the opposite edge of the patella facilitates the procedure. The needle can be directed below the patella or, if a large effusion is present, up and medially into the suprapatellar pouch. An anterior approach may be necessary when the joint is contracted in flexion. The needle is inserted on the medial or lateral side of the quadriceps tendon approximately 1–2 cm below the patella, with the knee in maximum flexion and the foot resting on the bed or couch.

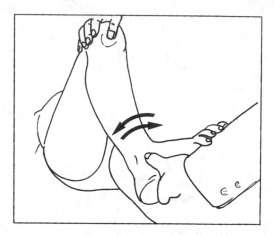

Figure 3–34. Testing for posterior lesions of the menisci. The knee is fully flexed and the thumb and index finger are placed along the joint line with the palm of the hand resting on the patella. The heel is then swept around in a U-shaped arc. The examiner may experience a click as the patient's face registers pain. This is best performed with the patient lying flat on his back.

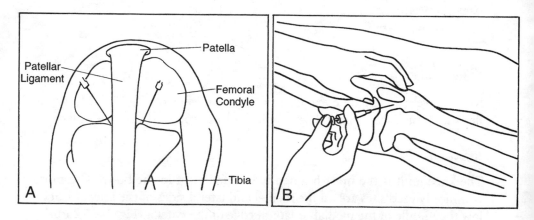

Figure 3–35. Technique of intraarticular injection of the knee. (A) Anterior approach (knee flexed) and (B) lateral approach. The left index finger pushes down the medial edge of the patella elevating the lateral edge. This maneuver increases joint space laterally. The medial and lateral approaches are most frequently performed. The anterior approach is useful when the knee is contracted in flexion.

ANKLE AND FOOT

The ankle and foot are the sites of frequent musculoskeletal complaints. Perhaps more than any other part of the musculoskeletal system, they tend to be overlooked by doctors. Examination of the ankles and feet, like other parts of the musculoskeletal system, should begin with inspection.

Inspection

When inspecting the ankle and foot, the patient's stance and gait should be noted by observing the patient when walking. The two phases in walking are the stance phase and the swing phase. The stance phase occurs when the foot is put on the ground. The swing phase includes the push off from the ground and the movement of the foot until it is placed down again (Fig. 3–36). When observing the stance phase, it should be determined whether the heel strikes the ground first, whether the patient has a flat foot, and how the foot is used to push off. It is also important to note the position of the foot when the heel and the first and fifth metatarsophalangeal joints are bearing the weight.

A painful heel will prevent the patient from striking the ground with the heel. This may occur in Reiter's disease, with periostitis of the calcaneum, and with calcaneal spur formation. When the heel is placed on the ground, the knee joint usually extends. If quadriceps muscle weakness or flexion contraction of the knee joint is present, extension does not occur. If there is a drop-foot due to paralysis of the dorsiflexors, the patient will characteristically "slap" the foot on the ground. During full weight bearing, pain arising in the metatarsophalangeal joints, metatarsal heads, plantar fascia, anterio-inferior aspect of the calcaneum, and the heel, will cause the patient to be unable to rest the foot flat on the ground. If quadriceps weakness is present, the knee will flex with weight bearing. During the push off, the patient will experience pain if there is disease of the metatarsophalangeal joints. When hallux rigidus is present, the patient often walks on the lateral aspects of the foot. Weakness of the gastrocnemius muscle results in the inability to push off and, therefore, the heel remains flat on the ground. Quadriceps weakness, if severe, often requires lateral rotation of the hip to swing the foot forward. When the dorsiflexors are weak, the foot tends to drag along the ground and must be lifted with flexion of the knee joint.

Inspection of the shoes is often as important as examination of the feet. Broken medial counters are due to flat feet and talar head prominence. The outer side of the heel is worn down in the normal foot. A valgus deformity of the rear foot wears down the medial side of the heel and distorts the counter and quarter. A drop-foot wears out the front of the sole. Equinovarus due to

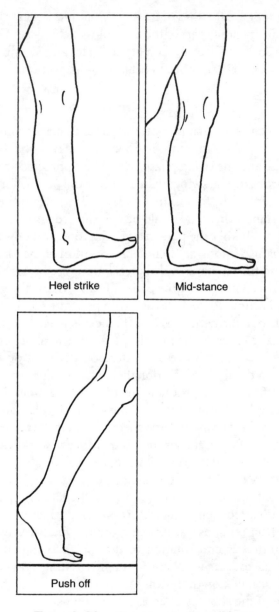

Figure 3–36. The phases in walking.

spastic paralysis wears out the front of the sole and the lateral border. Excessive wear on the lateral border of the sole also occurs with hallux rigidus. A deformed vamp with absence of creases indicates failure of plantar flexion of the foot, such as might occur with hallux rigidus or hammer toes. The wear mark across the sole indicates the position of the metatarsal heads.

Inspection of the foot should begin by noting the condition of the arches. The longitudinal arch is between the calcaneus and first metatarsal head, the highest point being the inferior border of the first cuneiform. The transverse arch lies under the first cuneiform and navicular bones medially, decreases laterally, and ends under the base of the fifth metatarsal and cuboid bones. The longitudinal and transverse arches disappear in flat feet. When flat feet are not rigid, the arches will be absent with weight bearing. Accentuation of the arches is seen in pes cavus, which is mostly due to neurologic causes, such as spina bifida, poliomyelitis, Friedreich's ataxia, and peroneal muscular atrophy. Pes adductus, or planus, is often seen in spastic children, while pes valgus is mostly due to flat foot. Pes equinus is usually the result of spastic paralysis, but can also occur in rheumatoid arthritis or trauma causing shortening of the Achilles tendon. Pes equinovarus is also known as club foot.

Skin alterations may indicate an underlying disease affecting the feet. Acute inflammation over the first metatarsophalangeal joint, especially if sudden in onset, is highly suggestive of acute gout. Subsequent peeling of the skin supports this diagnosis. Inflammation of the bursa overlying the lateral aspect of the first metatarsophalangeal joint is common in patients with hallux valgus. Skin changes, including pallor, cyanosis, skin atrophy, and gangrenous ulceration, may occur in atherosclerotic vessel disease. Impairment of vascular supply to the feet can be confirmed by palpation of the anterior and posterior tibial arteries and by pallor of the feet when the leg is elevated. When the feet are lowered from an elevated position, erythema is delayed by at least a minute and followed by intense erythema. Changes of progressive systemic sclerosis may also be seen in the feet. Calcium deposits may be palpated and may occasionally rupture and discharge through the skin. In progressive systemic sclerosis, Raynaud's phenomenon is common in the feet. Psoriatic skin and nail changes may occur in psoriatic arthritis. Keratoderma blenorrhagica may be seen in the soles in Reiter's disease. Diabetes causes vascular insufficiency and peripheral neuropathy, which may lead to diabetic gangrene. Diabetes also causes neuropathic joint disease, which most frequently affects the talonavicular joint.

In the early stages of rheumatoid arthritis, swelling and effusions into the metatarsophalangeal joints may cause the toes to be separated, giving the characteristic "window sign." In late stages of the disease, hallux valgus is common and the heads of the metatarsals may sublux. The skin becomes thick and inflamed over these subluxed heads and deep clefts occur in the soles between the metatarsal heads. The first toe may move laterally over the adjoining toes and the four lateral toes may become flexed (e.g., hammer toes). In rheumatoid arthritis, gangrene of the toes may occur. In psoriatic arthritis, sausage toes, similar to sausage fingers, may occur.

Palpation

The physician should be able to recognize the following bony anatomic landmarks; lateral and medial malleoli, calcaneum, sustentaculum tali, navicular bone, and metatarsal heads. Between the medial malleolus and the Achilles tendon are the posterior tibialis tendon, flexor digitorum longus, flexor hallucis longus, and posterior tibial artery and nerve. The spring ligament, which is frequently injured, is located between the sustenaculum tali and navicular bone. In the anterior compartment, which is located anteriorly between the two malleoli, from medial to lateral lie the tibialis anterior tendon, extensor hallucis longus, joint space, anterior tibial artery and nerve, and extensor digitorum longus. Around the lateral malleolus lie the peroneus longus and brevis, sural nerve, anterior and posterior talofibular ligament, calcaneofibular ligament, tarsal sinus, and origin of the extensor digitorum brevis.

Palpation of the undersurface of the feet is important. When periostitis is present, as in Reiter's syndrome, tenderness may be elicited over calcaneal spurs and the calcaneum. Calcaneal spurs are frequently found in asymptomatic patients. When plantar fasciitis and tenderness over spurs are present, it is preferable not to inject with corticosteroids. Rather, the patient should be provided with a heel "doughnut."

Tender heel pads are not uncommon in obese subjects who do a great deal of walking. Tenderness may be elicited behind the heel at the site of insertion of the Achilles tendon. Achilles tendinitis, due to inflammation of the paratenon (note the tendon has no surrounding tenosynovial layer), is common in athletes. Although the Achilles tendon is the strongest in the body, it often ruptures (completely or partially) 5 cm from its insertion. Injections of corticosteroid for treatment of Achilles tendinitis do not cause rupture. Patients receiving long-term oral corticosteroids who also have certain diseases, such as systemic lupus erythematosus, seem to be prone to rupture. The most common cause of rupture, in our experience, is old age. Thickening of the Achilles tendon may be due to chronic inflammatory changes in the parathenon, rheumatoid nodules, gouty tophi, or xanthomatous deposits.

Subachilles or retrocalcaneal bursitis may also cause pain and tenderness around the Achilles tendon. On palpation, the bursa is tender and bulges on either side of the tendon. This is not uncommon in patients with rheumatoid arthritis. These patients also develop an enthesitis at the insertion of the tendon, which causes erosion of the calcaneum (Fig. 4–10). More frequently, a subcutaneous bursitis is due to poorly fitting shoes. In boys between the ages of 10 and 15, pain and tenderness at the insertion of the Achilles tendon may be due to Sever's disease (retrocalcaneal apophysitis), which usually settles with restriction of activity and a heel pad. In polyarteritis nodosa, palpation of the sole of the foot may elicit small nodules due to aneurysms of the plantar arteries. Thickening of the plantar fascia, similar to that of the palmar fascia in Dupytren's contracture, may be palpated. This is known as Ledderhosen's disease and is a complication of diabetes mellitus.

Ligament sprains are common around the ankle, especially sprains of the anterior tibiofibular ligament. Sprains of the medial ligament are uncommon because the ligament is strong. When the medial ligament ruptures, widening of the medial joint space and diastasis of the inferior tibiofibular joint is evident on radiograph.

Tenderness in the midtarsal region in children may be due to Kohler's osteochondritis of the tarsal navicular. In adults, tender osteophytes on the dorsum of the talonavicular joint are common. Metatarsalgia is a general term for pain in the forefoot, especially around the metatarsal heads. Tenderness along the shaft of the second or third metatarsal bone may be due to a march fracture. Tenderness over the head of a single metatarsal in a child, especially the second metatarsal, is likely to be due to Freiberg's disease. Tenderness under the head of the first metatarsal may result from a stress fracture of one or both sesamoid bones in the flexor tendon. Palpation between the heads of the third and fourth metatarsals may reveal the site of a digital entrapment neuropathy (Morton's metatarsalgia). This is not to be confused with Morton's syndrome, in which a short first metatarsal causes excess weight to be borne by the second metatarsal head. This results in thickening of the shaft of the second metatarsal and hypermobility between the bases of the first and second metatarsals. A tender, rigid first metatarsophalangeal joint is usually the result of osteoarthritis (hallux rigidus). A hallux valgus is frequently complicated by inflammation of the overlying bursa (bunion). A congenital varus deformity of the fifth toe is relatively common, with the little toe crossing the fourth toe.

Inflammation of the bursa on the lateral aspect of the fifth metatarsal head commonly occurs in patients with rheumatoid arthritis and is often referred to as a "tailor's bunion." Pain around the base of the fifth metatarsophalangeal joint may also be due to peroneal brevis tendinitis. The heads of the metatarsals may become severely eroded in rheumatoid arthritis. In our experience, the first radiologic sign of erosion in rheumatoid arthritis occurs in the heads of the metatarsals. When the heads of the metatarsals sublux, they become tender. Deep clefts appear between the heads. The fat pads, which normally lie over the metatarsophalangeal joints, slip forward so that the patient has to bear weight with only calloused skin covering the joints. The toes develop "hammer" deformities (e.g., flexion contraction of the proximal interphalangeal joints). Hammer toes may also occur in patients who do not have rheumatoid arthritis and can become troublesome due to pressure of the shoe, producing a painful callosity over the dorsum of the flexed proximal interphalangeal joint. A "mallet" toe refers to a flexion deformity of the distal phalanx. Osteoarthritis can involve the interphalangeal joints of the toes.

Entrapment neuropathies involving the posterior tibial nerve (tarsal tunnel syndrome) and the medial plantar nerve are more common than previously realized. The posterior tibial nerve passes through a tunnel under the talocalcaneal or laciniate ligament, along with the tendons of the tibialis posterior, flexor digitorum longus, and flexor hallucis longus. The nerve has two branches. The medial branch supplies sensation to the heel and the medial three and a half toes. The lateral branch supplies sensation to the remainder of the sole and toes. Entrapment of the medial plantar nerve as it passes through an opening in the abductor hallucis muscle often occurs in patients with foot deformities (e.g., pes planus or pes cavus). Pressure over the anterior portion of the calcaneus reproduces the patient's symptoms. In addition to a Morton's neuroma, the digital nerves between the heads of the metatarsals may become compressed, causing loss of sensation to the adjacent surfaces of the toes. This most commonly occurs with the third plantar digital nerve. For further discussion of these and other entrapment syndromes, the reader is referred to the textbooks by Corrigan and Maitland and Pecina, et al. (3,10). The innervation of the foot is shown in Figure 3-37. The nerve roots and nerves controlling movement of the foot are listed in Table 3–2.

Special Diagnostic Signs

For anterior talofibular ligament insufficiency, the calcaneus is pulled anteriorly. For stability of the calcaneofibular and posterior talofibular ligaments, the calcaneus is inverted. On the lateral side, a gap appears between the talus and the calcaneus (the talocalcaneal gap). For deltoid ligament insufficiency, the calcaneum is everted. A gap between the talus and calcaneum indicates damage to the deltoid ligament.

Movements

Movements of the ankle take place in the talocrural and subtalar joints. The subtalar joints are not part of the ankle joint. Extension and flexion occur

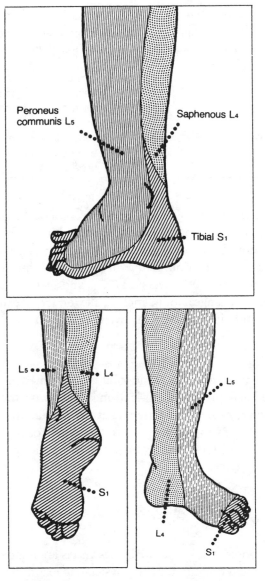

Figure 3–37.   Innervation of the foot.

Table 3–2. Nerve Roots, Peripheral Nerves, and Muscles Involved in Movement of the Foot

Movement	Nerve root	Peripheral nerve	Muscles
Plantar flexion	S1, S2	Tibial	Gastrocnemius and tibialis posterior
Dorsiflexion	L4, L5	Peroneal	Tibialis anterior, long extensors, peroneus tertius, and extensor digitorum brevis
Inversion	L4	Tibial and peroneal	Tibialis anterior and brevis, long extensors, and extensor digitorum brevis
Eversion	S1	Peroneal	Peroneal longus and brevis, long extensors, and extensor digitorum brevis

mostly in the talocrural joint. Eversion and inversion occur almost entirely in the subtalar joints. Inversion of the foot is a combination of flexion, adduction, and internal rotation. Eversion is a combination of extension (dorsiflexion), abduction, and external rotation. Range of movements include flexion (45°), extension or dorsiflexion (20°), inversion (35°), and eversion (25°). The movements of metatarsophalangeal joints include flexion (30°), extension or dorsiflexion (60–70°), and minimal lateral movements (abduction).

Injection

The approach to injection of the ankle joint is anteromedial or anterolateral. In the anteromedial approach, the injection is given with the foot in moderate plantar flexion. The needle is inserted at the medial aspect of the tibialis anterior tendon (Fig. 3–38). In the anterolateral approach, the needle is inserted just lateral to the tendon of extensor digitorum longus. This is the approach we favor.

SPINE

The spine is composed of 24 separate bones and is an anatomically complex structure. The spine is difficult for the physician to examine and radiology of the spine has limitations. CT and magnetic resonance imaging (MRI) are useful new methods of investigation of back pain. These methods are beginning to yield some of the secrets regarding the causes of back pain.

The detailed anatomy of the spine is well described in standard anatomy texts. The clinician should know about certain anatomic features in the spine

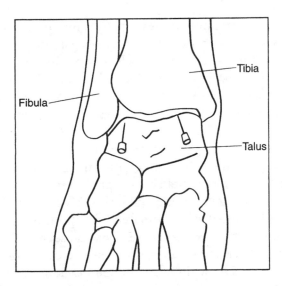

Figure 3–38. Technique of intraarticular injection of the ankle. Anteromedial and anterolateral approaches. Sites for injection of the ankle joint.

and be able to interpret radiographs of the atlantoaxial articulation. Details about interpreting radiographs are well described in standard textbooks of radiology and rheumatology. The clinician should know that the vertebral arteries and sympathetic plexus may be compressed or irritated as they pass through the foramen transversarium of C2 to C6. Osteophytes arising from the lateral portion of the corresponding disc margins may impinge on the vertebral arteries, as may osteophytes extending anteriorly from the uncovertebral joints of Lushka. Compression of the vertebral arteries may also occur as a result of posterior subluxation of the inferior facet joints. Although it is not well known, the vertebral arteries may be compressed by a cervical rib arising from the C7 vertebra. Flow in the vessels is compromised, especially in patients with atherosclerosis, and frequently occurs when the patient turns his head or looks upward. A cerebrovascular accident may occur with the clinical features of a posterior inferior cerebellar artery occlusion. The symptoms arising from vertebral artery compression are summarized in Table 3–3.

Table 3–3. Symptoms due to Vertebral Artery Insufficiency

Dizziness (1% only rotational), especially when turning the head or looking upward
Drop attack with or without loss of consciousness
Visual disturbance
Cerebral vascular incidences of posterior-inferior cerebellar artery nature

Dysphagia and perforaton of the esophagus may occur as a result of anterior osteophytes. In our experience, dysphagia occurs primarily with Forestier's disease (diffuse idiopathic skeletal hyperostosis).

The cervical discs begin between C2 and C3 and extend to between C7 and T1. The cervical nerve roots emerge at the upper border of the vertebra with the same number. The C3 root emerges between the C2 and C3 vertebrae. The additional root, C8, emerges between the C8 and T1 vertebrae. After this point, the roots emerge at the lower border of the vertebra of the same number. For example, the L3 root emerges at the L3–L4 disc space level. In the cervical spine, only one root can be compressed at the level of its exit. This is different in the lumbar spine and will be discussed later. The zygapophyseal joints of Lushka usually prevent lateral disc protrusion on the cervical spine. Inflammatory changes in soft tissues may cause cervical root irritation. Disc prolapse in the cervical spine is more likely to cause cord compression than root symptoms. The symptoms and neurologic signs of cervical nerve root compresson are summarized in Table 3–4. Headaches frequently are caused by degenerative cervical spinal disease.

Cervical Spine

Inspection

Inspection of the cervical spine begins when the patient enters the room. A torticollis may be immediately evident. There are two main causes of torticollis. The first cause is muscular spasm or contracture of the sternomastoid muscle, which may arise from a variety of conditions, such as cervical lymphadenopathy or acute cervical disc disease. As a result of this first cause, the head is displaced to the involved side. The second cause is muscular weakness, in which the head is displaced to the opposite, healthy side.

Although a flexed, stiff neck is often a feature of rheumatoid arthritis and ankylosing spondylitis, it can also occur in degenerative disease of the cervical discs and bilateral scalenus anticus syndrome. A "dropping" head is a feature of myasthenia gravis and myopathy. With severe involvement of the atlanto-occipital and atlanto-axial regions due to fracture, osteomyelitis, tuberculosis, rheumatoid arthritis, or neoplasm, the patient often holds her head when turning it (Rust sign).

A short neck with the hair margin at the level of the shoulders is characteristic of Klippel-Feil syndrome. A buffalo hump is found in corticosteroid-induced Cushing's syndrome and the less common spontaneous Cushing's syndrome. The buffalo hump consists of a mass of subcutaneous fat lying over the spinous processes of C7 and T1. In patients with diabetes mellitus, a corticosteroid-induced buffalo hump should not be confused with the

Table 3–4. Symptoms and Neurologic Signs Associated with Cervical Nerve Root Compression

Nerve root	Disc level	Symptoms	Weakness	Atrophy	Reflex change
C3	C2,C3	Pain and numbness at back of neck, especially around mastoid and pinna of ear	—	—	—
C4	C3,C4	Pain and numbness at back of neck, in levator scapulae, and rarely into anterior chest	—	—	—
C5	C4,C5	Pain radiating from side of neck to top of shoulder and numbness in distribution of axillary nerve	Abduction of arm and shoulder, especially beyond 90°	Deltoid muscle	Selective involvement of C5 segment causes an "inverted radial reflex."
C6	C5,C6	Pain down lateral side of arm and forearm (often into thumb and index fingers) and numbness on tip of thumb and dorsum of hand over first dorsal interosseous muscle	Biceps muscle	Biceps muscle	Biceps reflex

Table 3-4. Symptoms and Neurologic Signs Associated with Cervical Nerve Root Compression—*continued*

Nerve root	Disc level	Symptoms	Weakness	Atrophy	Reflex change
C7	C6,C7	Pain down middle of fore-arm (usually to middle finger) and, occasionally, index and ring fingers involved	Triceps muscle	Triceps muscle	Triceps reflex
C8	C7,T1	Pain down medial aspect of forearm to ring and small fingers; numbness in small finger and medial portion of ring finger; numbness rarely above the wrist	Triceps and small muscles of hand	Triceps and small muscles of hand	No reflex change

woody induration of Buschke's scleroedema in diabetes mellitus. A goiter may be present on the anterior aspect of the neck. Cervical lymphadenopathy may be visible and palpable in the posterior triangle and supraclavicular fossae. The most common cause of torticollis in a child is painful cervical lymphadenopathy. The salivary glands may be enlarged and submandibular salivary glands must be differentiated from enlarged submandibular lymph nodes. In Cushing's syndrome, the supraclavicular fossae may be filled out by fat pads. Emphysematous bullae may also cause swelling in the supraclavicular fossae. With fractured ribs or after operations on the thorax, subcutaneous emphysema may be present in the cervical region.

Although there are many causes of swelling in the neck, they are outside the scope of this book.

Palpation

The physician should know the following anatomical landmarks: occiput, inion, mastoid processes, and spinous processes of C7 and T1. The facet joints lie along lines through the mastoid processes parallel to the midline. Tenderness in these joints can be elicited by palpation. The patient should be lying supine with the head supported by the physician. The joint or joints that are tender can be determined by running the fingers along the lines of the facet joints. The advantage of this method is that the physician supports the patient's head, allowing the cervical muscles to relax.

An important aspect of the examination of the cervical spine is the diagnosis of atlantoaxial dislocation in rheumatoid arthritis, ankylosing spondylitis, and other inflammatory arthropathies. Pain is the most frequent symptom. It is felt in the occiput and, occasionally, radiates up the back of the head and over the vertex to be felt as a tight band around the forehead. The pain is dull and, when severe, may cause nausea. The patient may also experience symptoms resulting from vertebral artery and spinal cord compression. The second cervical vertebra may buckle anteriorly or posteriorly. If it buckles anteriorly, a bulge may be felt in the posterior pharyngeal wall. If it buckles posteriorly, the spine of C2 may be palpated when the thumb is drawn down from the occiput along the line of the cervical spinous processes. Instability of the head may occur and the patient may experience a "clunking" sound when the head is flexed and extended. This instability may be tested by rocking the head backward and forward. Neurologic complications of atlantoaxial dislocation are relatively uncommon, perhaps because the patient holds his neck stiff with little or no movement due to pain. When marked dislocation occurs, the spinal cord may be compressed. Basilar impression with the dens travelling up through the foramen magnum occurs primarily when the lateral masses of the first cervical vertebra collapse. The pain in the occiput is due to tension in the transverse ligament. Radiation over the skull follows the distribution of the second cervical nerve. Patients experience frontal and retro-orbital pain because there is often an anastomosis between the second cervical nerve and the ophthalmic branch of the trigeminal nerve.

The paravertebral muscles are difficult to examine, with the exception of the sternomastoid and trapezius. These muscles, especially the sternomastoid, may be in spasm or involved in a generalized muscle disease. The muscles are best examined in a relaxed position. Muscle power in the sternomastoids is easily tested by asking the patient to turn her head against resistance (Fig. 3–39) or lift her head from a resting position. Pressure on the greater occipital nerve may elicit pain in the area it supplies (Fig. 3–40). The carotid and subclavian pulses may be absent or severely diminished in Takayasu's arteritis involving the aortic arch.

Movements
The movements of the cervical spine can be tested in the sitting or recumbent position. When examining movement of the cervical spine, the examiner

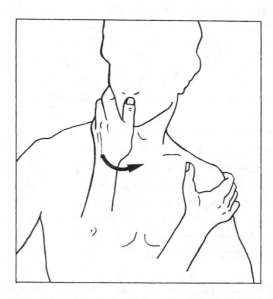

Figure 3–39. Testing muscle power in the sternomastoid by having the patient turn his head against resistance toward the suspected weak side.

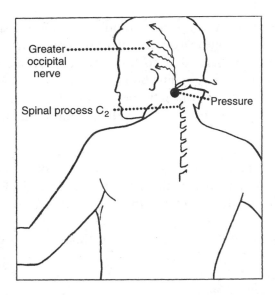

Figure 3–40. Pressure on the occipital nerve causes pain in the area supplied by the nerve.

should fix the trunk. Cervical spine disease frequently involves the cervical nerve roots.

If the physician suspects symptoms of cervical nerve root involvement, compression and elevation tests should be performed. Root pain may be provoked when the cervical spine is compressed by the physician pressing on the head or by lateral flexion of the head toward the affected side. Occasionally, pain may be elicited by lateral flexion.

When patients have symptoms in their arms and hands, rheumatologists and orthopaedic surgeons always palpate the supraclavicular regions for the presence of cervical ribs. Frequently, these are seen on routine chest radiographs and cause no symptoms. However, they may cause thoracic outlet syndrome and should always be considered in a patient with paraesthesia and vasomotor disturbance in the hand. Thoracic outlet syndrome usually appears in 20 to 40-year-old females of asthenic build. Pain and paresthesia occur in the distribution of the ulnar nerve in 95% of patients. Bruits may be heard over the subclavian artery. Horner's syndrome should suggest a superior sulcus pulmonary tumor or some other cause.

Clinical examination for evidence of thoracic outlet syndrome consists of two maneuvers. The first is to palpate for a cervical rib. The second is to examine for subclavian artery compression. This is done by having the patient take a deep breath and turn his head to the affected side (Fig. 3–41). When the occlusion is due to a cervical rib or scalenus anticus syndrome, the pulse at the wrist is obliterated. In costoclavicular syndrome, the pulse at the wrist may be obliterated by pulling both arms backward (Fig. 3–42). In hyperabduction syndrome, in which the pectoralis minor occludes the brachial artery, vein, and plexus, raising the arms obliterates the radial pulse (Fig. 3–43).

A Doppler study of the arterial pulsation and a nerve conduction study can further confirm the diagnosis of thoracic outlet syndrome. It is not important to determine whether the syndrome is caused by a cervical rib, scalenus anterior band, or narrow costoclavicular space, because each case requires resection of the first rib.

The many different causes of cervical pain and brachalgia are well described in two books by Dr. Rene Cailliet (13,14).

Thoracic and Lumbar Spine

Unlike the lumbar spine, disease of the thoracic spine is relatively uncommon. Disc prolapse rarely occurs, probably because the nucleus pulposus lies anteriorly in the disc space. The costovertebral joints may "slip" and cause

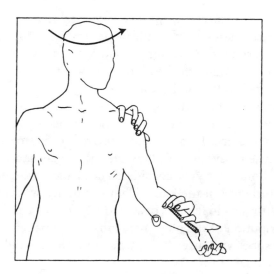

Figure 3–41. Adson's test. When the patient turns his head to the affected side, the radial pulse diminishes or disappears due to compression of the subclavian artery.

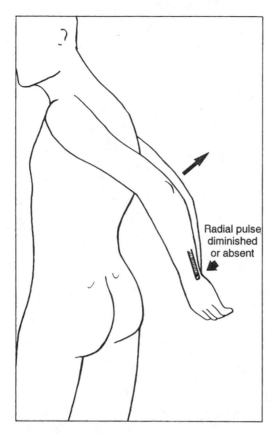

Figure 3–42. When both arms are pulled backward, the radial pulses diminish or disappear in the costoclavicular syndrome.

chest pain. Maximum movement of the spine occurs at the thoracic-lumbar junction. It is at this junction that a destructive discitis may occur in ankylosing spondylitis. Tuberculosis is now a rare disease affecting the thoracic spine. Osteoporosis and compression fractures are currently common in elderly women.

The lumbar spine is the site of modern man's Achilles heel. Approximately 65% of adults experience backaches at least once. Backache is second only to upper respiratory infections as a cause of absenteeism at work. It is frequently thought that problems with the lumbar spine began when man became erect. Yet, four-legged mammals suffer as frequently as man. The pressure in the nucleus pulposus of lumbar discs is highest, not when walking or running, but when sitting. This has been studied by Nachemson and Morris (15).

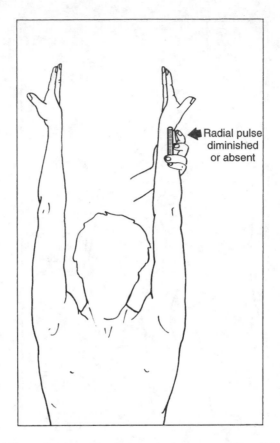

Figure 3–43.   When the hand is elevated, the pectoralis minor occludes the brachial artery, causing diminution or absence of the radial pulse.

The clinician should know that the spinal cord usually ends at the lower end of L1. This means that cord compression cannot occur below this level and that the lumbar sacral roots can be compressed at any level from L1 to their exit foramina. For example, a disc protruding between L5 and S1 may compress the L5 or S1 roots. If large enough, it may compress both roots. A large posterior central disc protrusion may cause compression of all the nerve roots in the cauda equina (Fig. 3–44).

The clinician should also know that an S1 root may be compressed at any point along its course from the L2–L3 disc level to its exit foramen. Therefore, a myelogram done to exclude the possibility of a disc protrusion involving the S1 nerve root is not complete until the dye reaches the L2–L3 disc level. A neuroma of the S1 root high in the cauda equina has been misdiag-

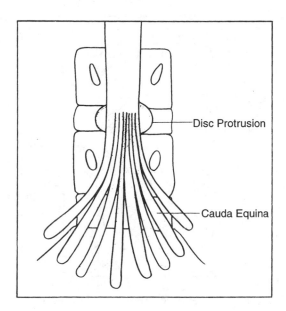

Figure 3–44. Central disc protrusion causes compression of the cauda equina. A disc protrusion may involve two nerve roots. It can be seen how a large posterior central disc protrusion may involve all the nerve roots in the cauda equina.

nosed as a slipped disc. However, this problem is rapidly becoming obsolete with MRI.

The articulations of the spine have three components: the disc and the two facet joints. The facet joints may be injured and may become involved with osteoarthritis or an inflammatory arthritis, such as ankylosing spondylitis. The facet joint syndrome includes lumbar and sciatic pain. Sciatic pain is the result of synovitis or periarticular inflammation, which may irritate and compress the nerve root (Fig. 3–45).

Most diagnoses of lumbago and sciatica can only be provisional, even with methods such as CT scans. To be certain about the cause of symptoms, it would be necessary to perform every relevant test on all of the patients. This is clearly impractical. Disease of the facet joint is indicated by acute short-lived attacks of lumbar pain, no increase in severity and duration with repeated attacks, occasional sciatica without neurologic deficit, and response to antiinflammatory analgesics.

The nucleus pulposus lies posteriorly in the discs of the lumbar spine. For this reason, posterior and posterolateral protrusion are common. Both the

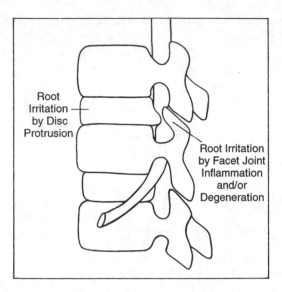

Figure 3–45. Root irritation and compression caused by disc protrusion or facet joint inflammation and degeneration. Root irritation can occur with disc protrusion and inflammation in or around the facet joint.

posterior aspect of the annulus fibrosus and the posterior longitudinal ligament are weaker than the anterior longitudinal ligament. Therefore, they are predisposing factors to a posterior rupture. The L5–S1 disc is thinner than other disc levels, wedge shaped, and the most common site of disc degeneration. Spondylolisthesis due to a defect in the pars interarticularis also commonly occurs at this site. The concept of the three joint complex in the spine is important because disc space narrowing, as a result of disc degeneration or surgery, results in abnormal stress in the facet joints. The facet joints then become unstable and develop osteoarthritis.

Because lumbar spine and root disease are so common, it is vital that the family practitioner know the symptoms and signs of lumbar and sacral root compression. These are summarized in Table 3–5. Omitted from this table are two important clinical entities: cauda equina syndrome and the syndrome of lumbar stenosis. Cauda equina syndrome is caused by a central posterior disc prolapse, large tumor, or other sources of compression of the nerve roots. The symptoms include bilateral sciatica, which may appear acutely or insidiously. Flacid paralysis, urinary retention, and impotence may ensue. The earliest bladder symptom is often dribbling incontinence. The anal (S2,S3,S4), bulbocavernous (S2,S3,S4), and cremasteric (L1,L2) reflexes are

Table 3–5. Neurologic Features of Lumbar and Sacral Nerve Root Compression

	Area of pain	Sensory loss	Motor deficit	Reflexes
L2	Across thigh	Usually none	Hip flexion	No loss
L3	Across thigh	Usually none	Knee extension	Adductor
L4	Down to medial malleolus	Medial leg	Inversion and dorsi-flexion of foot	Knee
L5	Posterior of thigh, antero-lateral calf, and medial aspect of foot and great toe	Dorsum of foot, medial aspect of foot and great toe	Dorsiflexion of foot and toes, especially big toe	No loss
S1	Posterior of thigh, thigh, posterior of calf, postero-lateral foot, and lateral toes	Behind lateral malleolus	Plantar flexion of foot and toes, eversion of foot, and atrophy of posterior compartment of leg	Ankle

lost. There is also perianal hypo- or anesthesia (S2,S3,S4). Therefore, all patients with lower back pain should be questioned regarding bladder symptoms. Rectal examination should include tests for perianal loss of sensation and loss of anal tone.

The syndrome of lumbar stenosis results from congenital narrowing of the lumbar canal or diseases that manifest in such narrowing (e.g., Paget's disease or diffuse idiopathic skeletal hyperostosis). The narrowing results in a diminished blood supply to the nerve roots, which can occur when walking. The patient complains of diffuse leg pains with walking, often associated with dysesthesia and paresthesia. The pains are relieved by sitting or adopting a supine position. Root pain may occasionally be associated with leg weakness and foot drop caused by walking. The syndrome is similar to intermittent claudication and is often referred to as spinal claudication. However, there is no absence of peripheral pulsation in the legs and pallor and rubor changes are not present. The patient often adopts a characteristic lordotic stance to relieve the pain.

Inspection
Inspection of the back may reveal café-au-lait spots. These immediately suggest neurofibromatosis or a "hairy spot," which may overlie a vertebral body abnormality. A painful back may be instantly diagnosed by the finding of a typical herpes zoster rash. Herpes zoster may be the presenting symptom of

a reticulosis or neoplasm infiltrating the spine. Psoriasis may appear as a single plaque in the lumbar region and may lead to the correct diagnosis of lower back pain due to sacroiliitis.

The posture and curvature of the spine are important in clinical examination. The normal curvature of the spine is shown in Figure 3-46. Many conditions may increase or decrease spinal curvature. Scoliosis may result from a number of different causes. The change in spinal curvature may be functional or organic. Functional alteration is reversible, whereas organic changes are not.

Lumbar lordosis is diminished in acute and chronic lumbar disc disease. In acute disc protrusion, the diminished lumbar lordosis is functional and improves when the condition settles. With chronic lumbar disc disease, the diminished lumbar lordosis may be irreversible (e.g., organic). Diminished lumbar lordosis is a common feature of ankylosing spondylitis. An increased

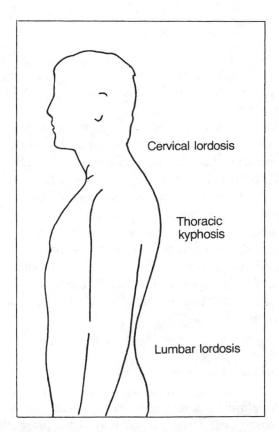

Figure 3–46.   Normal curvature of the spine.

lumbar lordosis is most frequently caused by obesity and weakness of the abdominal and paravertebral muscles. In congenital dislocation of the hip, especially when bilateral, marked increased lumbar lordosis may occur.

Kyphosis may occur as part of a scoliotic deformity. Localized kyphosis is seen in tuberculous disease of the spine and other conditions causing spinal destruction (e.g., trauma, osteoporosis, and neoplasms). In osteoporosis, collapse of a vertebral body tends to occur at the point of maximum movement of the spine (e.g., at the thoracic lumbar junction). Functional lumbar kyphosis may occur in acute lumbar disc protrusion and with abdominal pain (e.g., pancreatitis and pancreatic neoplasm, appendicitis). In addition to being organic or functional, scoliosis may also be compensated. This term refers to whether the scoliosis results from tilting of the pelvis due to real or apparent shortening of one leg. Therefore, when examining the patient for scoliosis, it is important to observe whether the iliac crests and the base of the sacral triangle are horizontal. The gluteal folds should normally be at the same level and the lengths of the legs should be equal. Functional scoliosis due to inequality of leg lengths may disappear when the patient sits. Flared lower ribs are often a sign of organic scoliosis.

The most common cause of functional lumbar scoliosis is an acute intervertebral disc protrusion. The direction of scoliosis does not indicate on which side the disc has prolapsed. A more useful sign is spasm of the paravertebral muscles, which overlie the site of disc protrusion. Although scoliosis and spasm of the paravertebral muscles may not be evident when the patient is standing, they become clear when the patient bends forward. It is not uncommon for a patient who has a small disc protrusion to develop a scoliosis when bent forward, which disappears on full flexion. This temporary scoliosis is easily recognized if the physician is aware of it.

Palpation

The physician should be aware of the following bony landmarks: spinous processes of vertebrae, iliac crests, posterior superior iliac spines, twelfth thoracic spine, lumbosacral junction, sacroiliac joints, first sacral foramina, sacrococcygeal junction, and coccyx.

Initial palpation of the spine should consist of the physician running her hand over the spinal processes. The patient should be sitting or standing with the back slightly flexed. Any alteration in contour can be ascertained (e.g., a "gibbus" or step-ladder effect), indicating a spondylolisthesis. Some physicians prefer to examine the patient lying prone, especially for percussion over the spinous processes.

Severe tenderness localized over one or two vertebrae on percussion should indicate the possibility of a local lesion, such as a collapsed vertebra in severe osteoporosis, tuberculosis, staphyloccal osteomyelitis, or secondary neoplasm. Pain in the spine may arise from injured or inflamed supraspinous and interspinous ligaments. In this case, the interspinous space is tender on percussion. The paravertebral muscles may be palpated for spasm and tenderness, and are best examined with the patient lying prone. Bilateral spasm of the paravertebral muscles is seen in disease affecting the vertebral bodies, such as staphylococcal osteomyelitis, spondylolisthesis, and ankylosing spondylitis. Unilateral spasm is more often a feature of acute disc prolapse. It is often useful to test whether bilateral paravertebral muscle spasm is voluntary or involuntary. This can easily be determined by asking the patient in the prone position to push his head downward on the bed. In a healthy subject, this movement of the head should result in tension of the abdominal muscles and relaxation of the paravertebral muscle. Tenderness in the paravertebral region that results in radicular pain is suggestive of nerve root compression, such as occurs in acute disc prolapse.

Palpation of the region between the fifth lumbar vertebra and the iliac crests may reveal tenderness due to inflammation. The inflammation may be due to long transverse processes of the fifth lumbar vertebra irritating the periosteum of the iliac crest or inflammation or strain of the iliolumbar ligament.

Palpation of the sacrum and sacroiliac joints may reveal spina bifida and the presence of tender, small, fatty Copeman herniae through the lumbodorsal fascia. Although there is no doubt that such herniae exist, there is controversy as to whether they cause symptoms. Subcutaneous nodules in rheumatoid arthritis may occur over the sacral region and have been described as the "forgotten" nodules. They may break down, ulcerate, and become a source of serious infection in a patient with rheumatoid arthritis who is confined to bed. The posterior superior iliac spine and the first sacral foramina may be tender in degenerative disc disease occurring between the fifth lumbar vertebra and the first sacral vertebra.

The sacroiliac joints are frequently involved in a number of seronegative polyarthritides. Tenderness can be elicited by direct palpation, percussion, and compression of the iliac wings (Fig. 3–47). Tenderness over the sacroiliac joints may also be due to strain or inflammation of the overlying posterior sacroiliac ligaments.

Pain in the coccygeal region is common. It may be due to local pathologic processes, such as injuries, rectal neoplasms, or referred pain arising from nerve root compression, especially from the fourth sacral root. In these instances, local tenderness of the coccyx may be present. However, it is important to perform a rectal examination and palpate the coccyx between the index finger and the thumb.

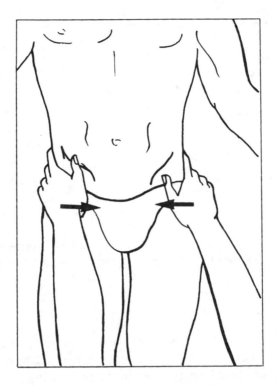

Figure 3–47. Sacroiliac tenderness can be elicited by compression of the iliac wings.

The gluteal musculature is the region most commonly used for intramuscular injections and, perhaps, the most common site in the body for iatrogenic abscesses. Spasm and wasting of gluteal muscles are best examined when the patient is lying prone. The sciatic nerve can be palpated halfway between the greater trochanter and the ischial tuberosity. When sciatica is present, the nerve is often tender and palpation may elicit pain in the distribution supplied by the nerve. The nerve is best palpated with the patient lying on her side with the hip and knee flexed (Fig. 3–48). Tenderness over the ischial tuberosity and the greater trochanter may be due to overlying bursitis or inflammation at tendinous insertions in ankylosing spondylitis.

Examination of the lumbar spine should include an examination of the abdomen. Intraabdominal disease may give rise to back pain. Likewise, lumbar spinal disease may cause abdominal pain. Recently, we saw a patient who had been investigated in orthopaedic and neurology departments and had a myelogram performed, among numerous other investigations, for lumbar

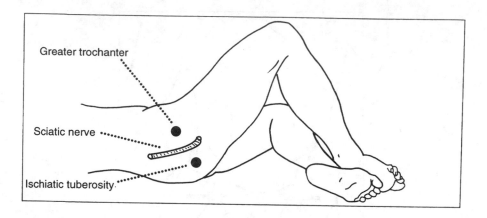

Figure 3–48. The sciatic nerve lies at the midpoint between the ischial tuberosity and the greater trochanter and can be palpated when the hip joint is flexed.

backache. A barium meal examination revealed that the patient had a posterior duodenal ulcer eroding the head of the pancreas. Examination of the abdominal muscles should be performed in the half-sitting position when the strength of contraction can easily be palpated. If there is unilateral weakness of the rectus abdominis, the umbilicus will be pulled toward the healthy side.

Movements

It is important to observe the patient bending forward. While doing so, the patient should have his knees extended. Lumbar scoliosis that develops when bending forward and disappears with full flexion is suggestive of an acute intervertebral disc prolapse. When bending forward, the distance between the spinous processes increases. If the separation is more than 3–4 cm, spinal flexion is probably normal (Fig. 3–49). Schöber's test is performed using a measuring tape to record distraction of skin marks on the lumbar spine (Fig. 3–50). Schöber's method has been modified by Macrae and Wright and corrected for age and sex (16,17,18). Recently, an arthrospinometer has been used for measuring spinal mobility and posture (19). The arthrospinometer will probably make the Schöber method obsolete. For a recent, extensive review see Bellamy (20).

Patients may not be able to touch their toes and still have normal spinal movement (21). Although obese subjects are prevented by abdominal enlargement, even thin subjects may be unable to touch their toes. Patients with

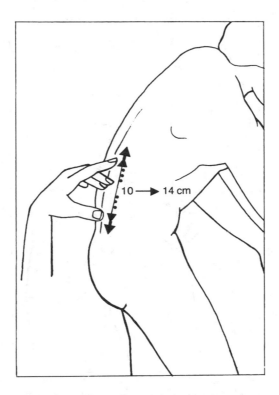

Figure 3–49. Simple method of measuring spinal movement. Distractor of more than 3–4 cm is probably normal.

ankylosing spondylitis and diminished spinal movement may be able to easily touch their toes due to good hip flexion and long arms.

Extension of the spine is also important to assess (Fig. 3–51). Although this is best accomplished with the patient lying prone, it can also be done with the patient in the sitting or standing position. It is best to anchor the spine by placing the palm of the hand over the posterior iliac spines. Extension of the spine is diminished in spinal disease and conditions affecting spinal musculature (e.g., polymyositis, myopathy, and myasthenia gravis). Extension of the spine should be tested actively and passively. When tested passively, weakness of spinal musculature may not be evident.

It is important to test lateral flexion when assessing spinal movement. The physician should first stabilize the pelvis before asking the patient to bend her trunk from side to side. Bilateral restriction in lateral movement of the spine occurs in a number of conditions, especially early ankylosing

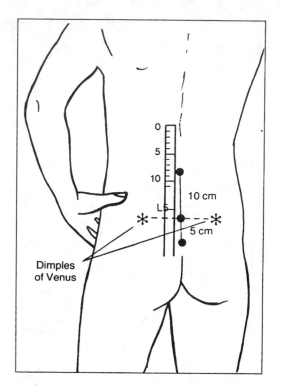

Figure 3–50. Points of measurement in modified Schöber's test. Approximately at the L5 spinous process, 10 cm above and 5 cm below the lumbosacral junction, marks are made on the skin when the patient is erect. The distance is then remeasured when the patient flexes forward. The L5 spinous process can be identified at the point of spinal intersection of a line joining the Dimples of Venus (marked as stars).

spondylitis. Unilateral restriction of lateral flexion is often seen in acute intervertebral disc prolapse. Lateral flexion may also provoke nerve root pain. To test rotation of the spine, the examiner should stabilize the pelvis, put the other hand on a shoulder, and turn the patient's trunk to the limit. The ability to rotate the spine is lost in the early stages of ankylosing spondylitis.

Neurologic signs may occur that do not quite fit the anatomic lesion. For instance, a prolapsed disc may give rise to two separate nerve root compression syndromes. For example, a prolapsed disc between the fifth lumbar and first sacral vertebrae may cause compression of the fifth lumbar and first sacral nerve roots.

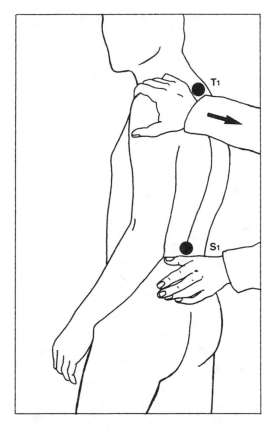

Figure 3–51. Extension of the spine assessed in the standing position.

Special Clinical Tests

Perhaps the most common test is the straight leg raising test of Lasègue (Fig. 3–52) (22). The leg should be raised to 70–80°. Pain arising from sciatic nerve irritation has a sharp, stabbing quality and tends to be localized in the segmental innervation of the sciatic nerve. The pain may be associated with paresthesia. Pain in the region of the nerve root compression may also occur. In this case, the patient frequently withdraws or turns in an attempt to lessen the pain. In the absence of nerve root compression, a positive Lasègue's test may occur with lumbar disc disease or facet joint involvement. The patient experiences pain in the lumbar region, which some physicians believe is more diffuse and can be differentiated from neuralgic pain occurring in nerve root compression. During Lasègue's test, patients may also experience pain due to taut hamstrings. This is easily differentiated from pain arising from the sciatic nerve.

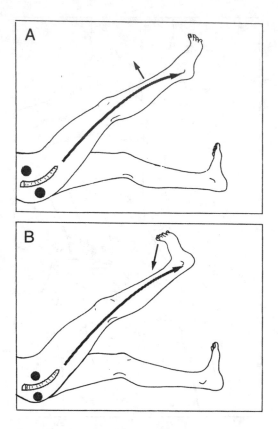

Figure 3–52.   Lasègue's sign. (A) The patient lies supine and the examiner raises the leg while the knee is kept straight. Pain may result from taut hamstring muscles, but is limited to the posterior thigh. (B) Sciatic pain can be provoked when the foot is dorsiflexed.

Ischial tuberosity pain usually accompanies pain caused by stretching of the hamstring muscles. A rare but important sign of a space occupying lesion at the level of the fourth lumbar vertebra is the crossed straight leg raising sign. When such a lesion is present, raising the leg on the opposite side of the disc protrusion causes radicular pain in the leg on the side of the protrusion. The patient often flexes the hip and knee to diminish the pain. When the leg opposite the side of protrusion is raised, the nerve root is pressed against the protrusion. Another useful sign is dorsiflexion of the foot during the straight leg test (21). This causes more severe pain as a result of stretching the sciatic nerve (Fig. 3–52). Dorsiflexion of the foot can also provoke pain as a result of stretching the calf muscles and pain in the calf if a deep venous thrombosis is present (Homan's sign). A positive Lasègue's sign can be confirmed by fixing the knees and asking the patient to sit up and bend forward. This provokes sciatic nerve pain. A negative Lasègue's test excludes a herniated disc in young patients, but not in those over 30. A good discussion of this test can be obtained in a review by Spangfort (19).

Another useful maneuver in performing Lasègue's test is known as the "bowstring" sign. When pain is elicited during the test, the knee is flexed. Finger pressure on the popliteal space will increase radicular pain. The sign is well described in the textbook by Macnab and McCulloch (21).

To test for third lumbar nerve root irritation, the femoral nerve is stretched. This can be done by having the patient lie on his side and, with the knee bent, pulling his leg backward. Pain down the front of the thigh will result.

Pain arising from the sacroiliac joints may radiate into the buttock and down the back of the thigh. In addition to direct palpation and pelvic compression, pain in the sacroiliac joints can be tested for by having the patient lie prone and elevating the leg with the examiner's hand on the sacroiliac jont. This will result in pain in the sacroiliac joint. This test is useful only when there is no hip joint disease.

Table 3–6. Common Causes of Back Pain

Children	Trauma
	Discitis
	Spondylitis (tuberculous or other)
	Osteomyelitis
	Meningismus
	Meningitis
	Epidural or subdural abscess
	Bleeding into spinal tissues
	Extramedullary tumors
	Renal disease
	Retroperitoneal tumors
	Intraspinal tumors
	Ankylosing spondylitis
	Sacroiliitis (tuberculous or other)
	Scheuermann's disease, rickets
Young adults	All of the above
	Discopathy
	Herniated discs
	Facet joint syndrome
	Myalgia
	Fibrosis
	Renal disease
	Gynecologic disease

The neurologic features of lumbar and sacral nerve root compression are summarized in Table 3–5 and the common causes of backache and sciatica in the different age groups are listed in Tables 3-6 through 3-8.

We do not believe that psychogenic backache is a relevant diagnosis. Many patients who are neurotic or depressed complain of backache and are

Table 3–7. Common Causes of Unilateral Sciatica

Root Involvement
   Disc protrusion
   Facet joint syndrome
   Spondylosis
   Spondylolisthesis
   Herpes Zoster
   Intraspinal tumors (e.g., neuroma, epidymoma, chordoma)
   Spinal disease (rheumatic, infectious, neoplastic)
   Epi- and subdural abscesses and hematomata
Sacral Plexus Involvement
   Retroperitoneal or pelvic pathology
     (e.g., tumors, inflammations, bleeding)
   Trauma
   Sacroiliitis
Peripheral Nerve Lesion
   Trauma
   Tunnel syndromes (e.g., piriformis syndrome, fibular head syndrome, tarsal tunnel syndrome)
   Infective, toxic, metabolic neuropathies (e.g., viral, alcoholic, diabetes)
   Neuroma
   Mononeuritis multiplex (rheumatoid arthritis, other forms of vasculitis)
   Peptic ulcer
   Other abdominal pathology, especially pancreatic
   Osteomalacia
   Hyperparathyroidism
   Spondylolisthesis
   Taylor's syndrome (epidural varicosity)
   Metastases
Middle Aged and Elderly
   All of the above
   Spondylosis
   Baastrup's syndrome
   Osteoporosis
   Paget's disease
   Chondrocalcinosis
   Vertebral hyperostosis
   Metastases
   Spinal stenosis

Table 3–8. Most Common Causes of Bilateral Sciatica

Cauda equina syndrome caused by tumor or centrally herniated disc
Spinal stenosis
Ankylosing spondylitis
Spondylolisthesis

difficult to treat. Neurosis or depression does not cause the backache. Bones and joints were the last to develop during evolution and totally changed the process of locomotion for the vertebrates. The spine and peripheral joints are not represented in the human psyche. The clinician must always seek a pathologic cause for backache. For example, we recently saw a patient with persistent thoracic spinal ache who was labelled as neurotic. On CT scan, the patient was found to have an epidural hemorrhage.

REFERENCES

1. Department of Health and Social Security. Morbidity Statistics from General Practice, Third National Study 1981–1982. England: Her Majesty's Stationary Office, 1986.
2. Cone R, Resnick D, Damzing L. Shoulder impingement syndrome: radiographic evaluation. Radiology 1984;150:29–33.
3. Corrigan B, Maitland GD. Practical orthopaedic medicine. London, England: Butterworth, 1983.
4. Hoppenfeld J. Physical examination of the spine and extremities. New York: Appleton-Century-Crofts, 1976.
5. Little H. The rheumatological physical examination. Orlando, Florida: Grune and Stratton, 1986.
6. Magee DJ. Orthopedic physical assessment. Philadelphia: WB Saunders, 1992.
7. Pecina MM, Krmpotic-Nemanic J, Markiewitz AD. Tunnel syndromes. Boca Raton, Florida: CRP Press, 1991.
8. Cailliet R. Hand pain and impairment, 3rd ed. Philadelphia: FA Davis Co., 1982.
9. Napier JR. The prehensile movements of the human hand. J Bone Jt Surg 1956;38B:902–913.
10. Docherty M, Hazleman BL, Hutton CW, et al. Rheumatology examination and injection techniques. London, England: WB Saunders Co., 1992.
11. Gerber JH, Dixon AS. Synovial cysts and juxta-articular bone cysts (geodes). Sem Arthritis Rheum 1974;3:223–248.
12. Schmitt BP, Kragg GR, Kelton JG. Acute calf pain and swelling in the arthritic patient: rupture of papulated cyst or deep venous thrombosis? Canad Med Assoc J 1981;125:54–56.
13. Cailliet R. Neck and arm pain. Philadelphia: FA Davis Co., 1981.
14. Cailliet R. Shoulder pain, 2nd ed. Philadelphia: FA Davis Co., 1981.
15. Nachemson A, Morris JM. In vitro measurement of intradiscal pressure: discometry, a method for determination of pressure in the lower lumbar discs. J Bone Surg 1964;46A:1077–1092.
16. Schöber von P. Lenden wirbelsaule und kreuzschmerzen. Munschr Med Wochenschrift 1937;9:336–338.

17. Macrae IF, Wright V. Measurement of back movement. Ann Rheum Dis 1969;28:584–589.
18. Moll JMH, Wright V. Normal range of spinal mobility: an objective clinical study. Ann Rheum Dis 1971;30:381–386.
19. Domjan L, Balint G. A new goniometer for measuring spinal and peripheral joint mobility. Hung Rheum 1987;28(suppl):71–76.
20. Bellamy N. Musculoskeletal clinical metrology. Kluwer, Dordrecht, 1993.
21. MacNab I, McCulloch J. Backache, 2nd ed. Baltimore: Williams & Wilkins, 1990.
22. Spangfort E. Lasègue's sign in patients with lumbar disc herniation. Acta Ortho 1971;42:459–461.

Radiology | 4

A practicing doctor should be able to read bone and joint radiographs and know when to consult a radiologist for a second opinion. The routine reporting of radiographs of bones and joints, which is currently practiced in hospitals, is time consuming and costly. Radiologists spend valuable time reporting details of fractures. Therefore, more education in radiology is needed for family and hospital doctors.

THE ABCs FOR EXAMINING RADIOGRAPHS OF BONES AND JOINTS IN BLACK AND WHITE (1,2,3)

It is important to know "The ABCs of Arthritis." These refer to alignment, bony mineralization, cartilage space, and soft tissue. The mnemonic was introduced by Forester, et al (4). It is useful in emphasizing the importance of a routine when examining a radiograph of joints. The routine that we favor includes a general inspection with inspection of soft tissue, cartilage, and bone.

General Inspection

The radiograph should be checked for film density. Essentially, black should be black and white should be white. When placed behind the black part of the film, the fingers should not be seen. Artifacts, such as plasters and buttons, should be noted.

Soft Tissue

Radiographs should be examined for swelling of the soft tissues, injuries, scars, and ulceration. Calcification in the soft tissues should immediately suggest progressive systemic sclerosis and dermatomyositis. Tophi may be readily diagnosed. Foreign bodies (e.g., broken-off ends of needles, bullets) should be noted. Effusions in joints can be easily recognized. Opacities in inflamed bursae (e.g., tendon Achilles bursitis) and thickening of tendons may also be seen. Baker's cysts behind the knee joint appear as opaque masses. Synovial hypertrophy may cause a bulge from the margin of the joint and is often seen on the anterior aspect of the ankle joint. Doctors Weston and Palmer published an excellent book on soft tissue changes that may be visualized on straight radiographs (5). Xeroradiography is also useful in identifying soft tissue lesions, although it is seldom used today.

Cartilage

Although cartilage cannot be seen on a plain radiograph of the joint, loss of cartilage causes loss of joint space. However, cartilage may be absent in the joint when there is no loss of joint space. This occurs when the space has been replaced by inflamed synovium (e.g., tuberculous synovitis in the hip). Calcification of cartilage (chondrocalcinosis) is due to the deposition of salts of calcium pyrophosphate.

Bone

When initially examining bone, density should be considered. There are many causes of osteopenia (Table 4-1). The most common causes are postmenopausal and senile osteoporosis. Osteopenia refers to "poverty of bone" and can be due to osteoporosis or osteomalacia. Osteoporosis refers to reduction of bone and matrix in equal proportion, while osteomalacia is a deficiency in mineralization due to reduction of calcium phosphate levels below that required for mineralization of matrix. The characteristic features of osteomalacia are pseudofractures or a Looser's zone. These consist of radiolucent bands of osteoid, probably representing partial insufficiency fractures, lying at right angles to the surface of the bone. They are principally seen in the femoral neck, pubic rami, long bones, ribs, metatarsals, and scapulae. Although they may persist unchanged for months or years, true fractures may occur as a result of them due to weakening of the bone. Rickets, the childhood form of osteomalacia, is characterized by widening of the growth plate and flaring of the metaphyses.

Table 4–1.   Causes of Osteopenia

Osteoporosis
1. Most common. Senile and postmenopausal osteoporosis. Senile type also occurs in progeria.
2. Generalized systemic disease. Rheumatoid arthritis and other inflammatory arthritides, chronic obstructive pulmonary disease, diffuse bone metastases, and amyloidosis.
3. Endocrine disease. Hyperthyroidism, hyperparathyroidism, Cushing's disease, acromegaly, diabetes mellitus Type I, hypopituitarism, hypogonadism (e.g., Klinefelter's syndrome) and premature ovarian failure, pseudohypoparathyroidism, and athletic amenorrhea.
4. Hematologic diseases. Sickle cell anemia, thalassanemia, hemophilia and Christmas disease, hemochromatosis, and multiple myeloma.
5. Genetic disorders. Osteogenesis imperfecta, Marfan's syndrome, homocystinuria, mucopolysaccharidosis, Gaucher's disease, lipoidosis, ochronosis, glycogen storage disease, and Turner's syndrome.
6. Immobilization. Hemiplegia, paraplegia, and weightlessness.
7. Dietary deficiency and eating disorders. Protein and/or calcium deficiency, scurvy, anorexia nervosa, and bulimia.
8. Iatrogenic. Corticosteroid therapy, chronic alcoholism, prolonged heparin therapy (more than 15,000 units/day), gonadotrophin releasing agonists and antagonists, and treatment with thyroxine, especially for thyroid cancer.

Osteomalacia
1. Reduced vitamin D intake. Malabsorption [post gastrectomy, celiac disease, idiopathic steatorrhea, Crohn's disease, hepatobiliary disease (especially chronic biliary cirrhosis), and pancreatic insufficiency]. Nutritional (e.g., total parenteral nutrition and lack of exposure to sunlight, which is rare.
2. Defective metabolism of vitamin D. Chronic renal failure, phenobarbitone and dilantin therapy for epilepsy, hepatobiliary disease, X-linked hypophosphatemia, Vitamin D dependent rickets Type I (Type II is a receptor defect), and oncogenic hypophosphatemic osteomalacia.
3. Renal phosphate loss. X-linked hypophosphatemia (familial hypophosphate rickets), Fanconi syndromes (Wilson disease, cystinosis, oculocerebral syndrome, tyrosinemia, paraproteinemias, monoclonal gammopathies, Sjögren's syndrome, systemic lupus erythematosus, and heavy metal poisoning), and oncogenic hypophosphatemic osteomalacia.
4. Chronic phosphate malabsorption, antacid abuse, and malabsorption.
5. Defective mineralization. Aluminum toxicity, fluorosis, and treatment with diphosphonates (sodium etidronate).

Osteoporosis results in loss of compact cortical bone and cancellous or trabecular bone. Although the skeleton is comprised of 80% of compact cortical bone, the turnover in cancellous bone is eight times more rapid. Therefore, osteoporosis and any cause of osteopenia will first become evident in

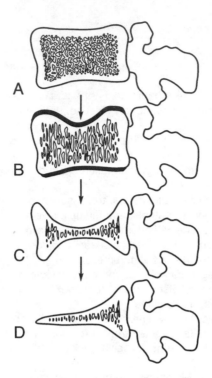

Figure 4-1.   Successive changes in the spine in osteoporosis. Note the vertical streaking, apparent increased density of vertebral endplates, and cortical thinning of the body of the vertebra (B) compared to the normal spine (A). Note how the vertebral body becomes empty of trabeculae and the discs protrude into the body, giving the "codfish" or "fish mouth" appearance (C). Finally, there is collapse of the body (D).

cancellous bone. Bone loss is first evident in the spine and is apparent in straight radiographs after 30% of bone mineral is lost. The body of the vertebrae primarily consist of cancellous bone, with compact bone only present in the vertebral end plates and posterior osseous elements. The changes seen in the spine in osteoporosis are illustrated in Figure 4-1. Resorption of horizontal trabeculae results in the appearance of vertical streaking. Thinning of cortical bone becomes evident and is especially seen on the anterior border of the vertebrae. Although sclerosis does not increase, there is an apparent increase in density of the vertebral end plates. The vertical trabeculae are resorbed, resulting in a "picture frame" or "empty box" appearance. The intervertebral discs may protrude into the end plates, causing buckling with a "codfish" or

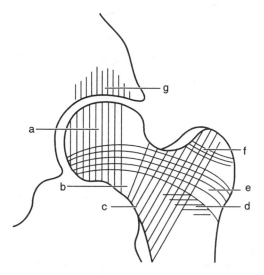

Figure 4-2. The groups of trabeculae in the femoral head and neck, which form the basis of the Singh index. There are two principal groups and four secondary groups. Note Ward's Triangle (b). The two principal groups of trabeculae are the principal compressive groups (a) and the principal tensile groups (e). The secondary groups include the secondary lateral compressive trabeculae (c), secondary tensile group (d), greater trochanter group (f), and trabeculae in the acetabulum (g).

"fish mouth" appearance. The nucleus pulposus may burst through the vertebral end plate, giving rise to Schmorl's nodes. Collapse of vertebral bodies may result, causing kyphosis or a "dowager's lump." Osteopenia can also be easily recognized by loss of the trabecula pattern of the proximal end of the femur. The five major groups of trabeculae are illustrated in Figure 4-2.

When examining standard radiographs of bones, it is important to focus on radiolucency and thickness of the cortex of bone (Fig. 4-3). When rapid loss of bone occurs, as in reflex sympathetic dystrophy, slight subperiosteal resorption may be seen, especially with magnification views. A moderate or severe degree of subperiosteal resorption of bone is pathognomonic of hyperparathyroidism. Although these changes in cortical bone are never seen in postmenopausal or senile osteoporosis, they may be seen in a variety of conditions, including hyperparathyroidism, renal osteodystrophy, acromegaly, osteomalacia, reflex sympathetic dystrophy, and disuse osteoporosis. There are several causes of localized osteoporosis (Table 4-2) and focal and diffuse osteosclerosis (Table 4-3).

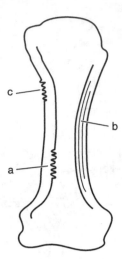

Figure 4-3. Sites of osseous resorption in cortical bone in osteoporosis: (a) endosteal, (b) intracortical appearing as longitudinal striations, and (c) very mild subperiosteal resorption.

The family doctor is more likely to encounter metastases in bone than primary tumors of bone, which are rare. Some tumors are more likely to metastasize to bone than others. For example, osseous metastases can be found at autopsy in 75% of patients with breast cancer, while they are only found in 10% of patients with gastrointestinal malignant tumors. Metastases of bone may be lytic or sclerotic. The metastases from some tumors (e.g., breast carcinoma) tend to be lytic, while others (e.g., prostate carcinoma) tend to be sclerotic. Metastases may become sclerotic after chemotherapy or radiotherapy. Metastases of the hand or foot often originate from a bronchogenic carcinoma. A solitary expansile lytic metastasis usually arises from the thyroid or kidney, rarely from a melanoma or pheochromocytoma.

Table 4–2   Causes of Localized Osteoporosis

Immobilization (cast, hemiplegia, and paraplegia)
Reflex sympathetic dystrophy (Sudeck's atrophy)
Transient regional osteoporosis (especially in the hip during pregnancy, and also a migratory type)
Paget's disease (active phase)
Inflammatory joint disease (rheumatoid arthritis, seronegative spondyloarthropathies, septic arthritis and tuberculosis—all juxta-articular)

Table 4–3.   Causes of Osteosclerosis

1. Healing Lesions
   Stress fractures
   Bone infarcts
   Brown's tumors in hyperparathyroidism and osteolytic metastases after chemother-
      apy and radiation
2. Tumors
   Metastases (especially from prostate, osteoma, osteoid osteoma, and os-
      teoblastoma)
   Primary bone sarcoma
   Lymphoma, leukemia, mastocytosis, and multiple myeloma (1–3% are sclerotic)
3. Infections
   Osteomyelitis, especially the sclerosing type of Garré; Brodie's abscess if involv-
      ing the cortex; and syphilis (e.g., saber tibia)
4. Miscellaneous
   Paget's disease
   Myelofibrosis (often preceded by osteopenia in the early stages)
   Renal osteodystrophy and hyperparathyroidism, especially secondary (e.g.,
      "rugger-jersey" spine)
   Fibrosis
   Vitamin D intoxication in children, often followed by osteoporosis
   Vitamin A intoxication in children (but not before the age of 6 months)
   Bismuth and lead poisoning (widened metaphyses with dense bands)
   Gaucher's disease (in reparative stage)
   Fibrous dysplasia
   Osteopetrosis (Albers-Schonberg disease)
   Engelmann's disease (progressive diaphyseal dysplasia)
   Caffey's disease (infantile cortical hyperostosis)
   Osteopoikilosis
   Osteopathia striata (Voorhoeve's disease)
   Tuberose sclerosis

The differentiation of benign (nonaggressive) and malignant (aggres-
sive) tumors of bone is determined according to the age of the patient (Ewing's sarcoma is rare after 20), whether pain is associated with the tumor
(more likely to be malignant), whether there is any soft tissue swelling or tenderness, and whether there are systemic symptoms or signs, such as elevated
erythrocyte sedimentation rate (more likely to be malignant). A solitary lesion is more likely to be benign, while multiple lesions tend to be malignant.
The site of the tumor often helps to identify its nature (Fig. 4-4), as do other
features (Table 4-4). Further investigations are required, such as radioisotope
bone scans, computed tomography (CT), and magnetic resonance imaging
(MRI). (See "Special Investigations".)

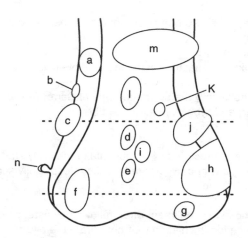

Figure 4-4. Sites of primary tumors of bone. The interrupted horizontal lines separate the diaphysis above from the metaphysis in the middle and epiphysis below. Note that in the child, a giant cell tumor is metaphyseal, but in the adult often begins in the epiphysis. (Modified with permission from: Madewell JE, Ragsdale BD, Sweet DE. Radiologic and pathologic analysis of solitary bone lesions. Part I: Internal margins. Radiol Clin N Am 1981;19:715–748; and Chapman S, Nakielny R. Aids to radiological differential diagnosis. London, England: Baillère Tindall, 1984).

   Key: Adamantinoma (a), Osteoid osteoma (b), chondromyxoid fibroma (c), simple bone cyst (d), osteochondroma (e), giant cell tumors (f), enchondroma and chondrosarcoma (g), chondrosarcoma (h), osteosarcoma (i), aneurysmal bone cyst (j), nonossifying fibroma (k), fibrosarcoma (l), fibrous dysplasia (l) and reticulum cell sarcoma, Ewing's tumor and chondrosarcoma (m), and osteochondroma (n).

Table 4-4.   Radiographic Features that may help to Differentiate Benign from Malignant Primary Tumors of Bone

	Benign	Malignant
Size	Usually small	Often large
Soft tissue mass	None	Common, often tender and inflamed
Margins	Sharply outlined; sclerotic with reactive bone formation	Poorly defined or no margins; no reactive bone formation
Transition zone	Sharply defined narrow zone between lesion and normal bone	Ill-defined wide zone of transition due to aggressive pattern of destruction
Periosteal reaction	Rare; if present uninterrupted solid periosteal reaction, which contains the lesion	Common with "sunburst" pattern with Codman's triangle in osteogenic sarcoma and "onionskin" or laminated periosteal reaction in Ewing's sarcoma
Overall appearance	Geographic uniform destruction of bone	Moth-eaten, ragged borders, which are ill-defined and permeative
New bone	None	Common, often irregular
Calcification	Often present, but clearly defined	Often present, but irregular and "cloud-like" or "cotton wool" in appearance
Systemic features	None	Frequent, with high sedimentation rate

EVERYDAY ARTHRITIS

Osteoarthritis

Osteoarthritis is characterized radiologically by four essential features (Fig. 4-5). These include loss of joint space, subchondral new bone formation, juxta-articular cysts formation, and formation of osteophytes.

The first three features are due to destructive changes, while the fourth feature is reparative. Loss of joint space is due to erosion of cartilage, while subchondral new bone formation is the result of increased vascularity. The bony sclerosis may be so severe that it resembles ivory; hence, the term eburnation. Juxta-articular bone cysts, which do not have an epithelial lining, are the result of synovial fluid intrusion through the damaged cartilage or cartilage-denuded bone, cystic necrosis due to bony contusion from impacting

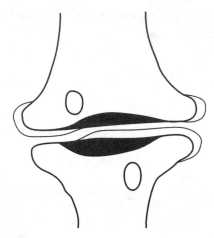

Figure 4-5. Osteoarthritis in the distal interphalangeal joint (Heberden's node). Note loss of joint space due to attrition of cartilage, osteosclerosis (eburnation) of juxta-articular bone, bone cysts, and osteophytes capped with cartilage.

osseous surfaces, or a combination of both. These cysts are often referred to by the French word, géode. Joints demonstrating these cysts do not appear to be more painful than joints without cysts. Often, an early sign of osteoarthritis of the hip is the development of a bone cyst (Egger's cyst) in the acetabulum. Osteophytes primarily occur at joint margins and are the reason why the disease was originally called hypertrophic arthritis.

Swelling of Heberden's nodes cannot be explained by osteophyte formation alone, because the osteophytes are frequently small. The swelling is primarily the result of a "cap" of articular cartilage on the osteophyte (Fig. 4-5). The interphalangeal joint of the thumb is frequently involved in osteoarthritis. In 10% of patients, osteoarthritis is also present in the metacarpophalangeal joints. Osteoarthritis of the first carpometacarpal (trapeziometacarpal) and trapezioscaphoid joints is a frequent cause of severe pain and disability. If only the trapezioscaphoid joints are affected, chondrocalcinosis may be implicated. Confusion often arises with the names of the carpal bones. The trapezium was originally the greater multangular and the scaphoid was originally the navicular. Although osteoarthritis is essentially a nonerosive disease, an erosive form exists and has the same distribution in the joints of the hands as the ordinary form of osteoarthritis. If the erosion is present in the middle of the joint, a "seagull" appearance results. The marginal erosions in psoriatic arthritis resemble "mouse ears." Due to the considerable amount of

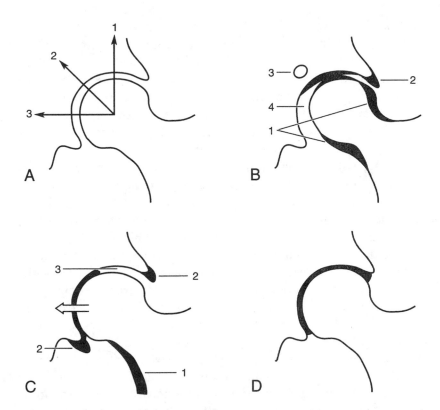

Figure 4-6. (A) Normal hip joint with patterns of migration: (1) superior, (2) axial, and (3) medial. (B) Most common form of primary osteoarthritis (80% of patients) affecting the superior and lateral aspect of the joint. Note (1) femoral neck buttressing, (2) osteophyte on lateral portion of the acetabulum, (3) Egger's cyst in the acetabulum, and (4) widening of the inferomedial aspect of the joint. (C) Second form of primary osteoarthritis (20% of patients). Note (1) buttressing of the femoral neck, (2) osteophyte formation, (3) widening of the superolateral aspect of the joint, and (4) medial migration, which produces a mild protrusio acetabuli. (D) Symmetrical loss of joint space in secondary osteoarthritis (see text for causes). Note the absence of osteophytes.

synovitis present in erosive osteoarthritis, it is not surprising that bony ankylosis can occur. However, ankylosis of joints is not a feature of osteoarthritis.

Osteoarthritis is most disabling when it affects the hips and knees. When the joint space begins to narrow, the femoral head migrates upward or medially (Fig. 4-6). Migration upward is often unilateral, while medial migration is usually bilateral. Both these patterns of loss of joint space associated with osteoarthritis are nonuniform, more frequent in women, and considered examples of primary osteoarthritis. Thickening or "buttressing" of the femoral neck occurs particularly on the medial side. Mild to moderate protrusio acetabuli may develop in the medial migration pattern. Osteophytes may develop at the margins of the fovea and beneath the reflected synovium around the femoral neck, where they may be mistaken for a femoral neck fracture. Osteophyte lipping of the lateral aspect of the femoral neck can sometimes produce cyst-like radiolucent areas.

A third type of axial migration of the femoral head occurs due to diffuse or concentric loss of cartilage. This is relatively rare and often secondary to another disease (e.g., rheumatoid arthritis, septic arthritis, calcium pyrophosphate dihydrate crystal deposition disease, or ochronotic arthropathy). Protrusio deformity is common in this type of hip arthroplasty.

The performance of hip arthroplasties is complicated by the quality of the patient's bones. Increase in bone mass in all joints is associated with osteoarthritis. A decrease in bone mass is associated with fracture of the femoral neck and other bony fractures. Therefore, although osteoporosis and osteoarthritis are both common in elderly patients, there is an inverse correlation between them.

The knee joint is commonly involved in osteoarthritis and is the most common cause of physical disability in elderly people. The knee joint consists of three compartments, including the medial femorotibial, lateral femorotibial, and patellofemoral compartments. Although osteoarthritis may affect all three compartments, the medial aspect of the joint is the most frequently involved. When the disease affects the lateral femorotibial compartment, it is frequently the result of injury, such as laxity of the cruciate ligament. Radiographs of the knees in a patient with osteoarthritis should be taken while the patient is weight bearing with the knees slightly flexed (standing tunnel projection). This position results in better recognition of cartilage loss. The tibial spines are often prominent and sharpened in early osteoarthritis. Intraarticular osteophytes can be mistaken for osseous loose bodies and "joint mice," both of which are common in severe disease. Osteoarthritis affecting the patellofemoral compartment is best visualized on lateral films with the knee slightly flexed. Anterior scalloping of the lower end of the femur may result from pressure of the patella with the knee extended.

Although primary osteoarthritis rarely affects the ankle, it commonly affects the tarsal joints and first metatarsophalangeal joint (hallus rigidus). The metatarsophalangeal joint is frequently associated with hallux valgus, when the sesamoid bones can be seen to be displaced laterally.

Osteoarthritis can affect the shoulder joint and acromioclavicular joint. It is frequently associated with a shoulder impingement syndrome (e.g., en-

Table 4–5. Causes of Neuropathic or Charcot's Arthropathy

Intraarticular corticosteroid injections	Spinal cord injury
Congenital insensitivity to pain	Multiple sclerosis
Charcot-Marie Tooth disease	Leprosy
Dysautonomia (Riley-Day syndrome)	Alcoholism
Diabetes mellitus	Amyloidosis
Tabes dorsalis	Syringomyelia

croachment of the subacromial space by upward displacement of the humerus). Osteophytes may be seen in the area of the bicipital groove, which may be due to ossification of the sleeve of the long head of biceps. The sternoclavicular joint, elbow joints, and sacroiliac joint (particularly on the iliac side) may also be involved in osteoarthritis. Osteophytes, especially in the superior and inferior portions of the joint, may overlie the interosseous space and be misinterpreted as ankylosing spondylitis. In contrast to the irregular outline in ankylosing spondylitis, the adjacent subchondral bone is well defined. Bridging osteophytes are sometimes seen in elderly patients at the anterior-inferior aspect of the joint. Although not an "everyday arthritis," neuropathic or Charcot's arthropathy may be seen, especially by endocrinologists in diabetic clinics. There are many causes of neuropathic arthritis (Table 4-5), which can be viewed as a form of osteoarthritis "with a vengeance." There is often complete dissolution of the joint, subchondral sclerosis, and fragments of bone within and outside of the joints. Massive new bone and osteophyte formation are common. There may be extensive bone resorption, giving the appearance of surgical amputation. The radiologic distinction between a neuropathic joint and septic arthritis is that a neuropathic joint shows juxta-articular osteopenia in contrast to the increased sclerosis of septic arthritis.

Rheumatoid Arthritis

The radiologic changes associated with this disease can best be understood by recognizing the sequence of pathologic changes that may occur (Fig. 4-7). The earliest changes on radiographs are soft tissue swelling and juxta-articular osteoporosis. The soft tissue swelling is due to edema of the soft tissues around the joint and joint effusion. Although juxta-articular osteoporosis is obvious, its interpretation has a high intra- and interobserver error. Erosions are first apparent in the bare areas of bone, such as bone within the joint capsule that is not covered by articular cartilage and is in direct contact with synovium (Fig. 4-7). Initially, the erosive changes have a "dot-dash" appearance, which is best seen with magnification (Fig. 4-8). This is followed by a definite

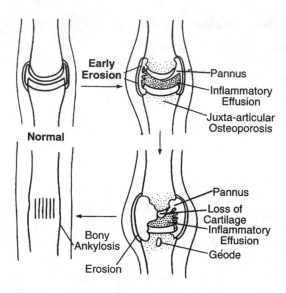

Figure 4-7. Progression of pathologic changes in rheumatoid arthritis.

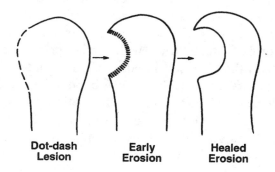

Figure 4-8. Radiologic stages of marginal erosion in rheumatoid arthritis.

loss of bone with a soft "cotton wool" base. When the erosion heals there is a hard, firm base. As the hypertrophied inflamed synovium (the pannus) grows over the surface of the cartilage and between the cartilage and bone, there is progressive loss of cartilage with reduction in joint space. Juxta-articular bone cysts develop similar to those seen in osteoarthritis. (However, in osteoarthritis, the margins of the cysts are well defined and sclerotic.) In severe cases, the joint may be completely destroyed, giving rise to arthritis mutilans. In the hands, arthritis mutilans may be described as "opera glass" hand or "main en lorgnette." In the late stages of the disease, fibrous and bony ankylosis may occur. Bony ankylosis is defined when cancellous bone flows without interruption from one bone to another with no evidence of joint space. In our experience, bony ankylosis occurs most commonly in the wrists of patients who consistently test negative for serum IgM rheumatoid factor.

In addition to juxta-articular osteoporosis, generalized osteoporosis also occurs, especially in postmenopausal female patients with disease of long duration and marked severity. Oral corticosteroid therapy also contributes to the development of generalized osteoporosis. In patients with rheumatoid arthritis, radionuclide studies demonstrate a generalized increase in metabolic activity of bone with increase in bone turnover. Osteoporosis predisposes patients to fractures, especially crush fractures of the vertebrae. A mild periostitis may be seen near active joints. Unlike psoriatic arthritis and Reiter's disease, periostitis is never florid in rheumatoid arthritis.

In early rheumatoid arthritis, the clinician should know the areas to examine for articular erosions. In a study of 100 patients with early rheumatoid arthritis (e.g., less than one year with the disease) the feet were the most common site of erosive disease. In particular, erosions were seen in the heads of the second and fifth metatarsals. In the hands and wrists, early erosion was detected in the second and third metacarpophalangeal joints, ulnar styloid, and carpal joints. Erosions of the ulnar styloid are due to synovitis in the prestyloid recess of the radiocarpal compartment and inferior radioulnar compartment. Surface erosions are also seen on the lateral aspect of the ulnar styloid and are probably due to synovitis of the extensor carpi ulnaris tendon sheath. Erosive changes in the carpal joints are more apparent in the dominant hand.

All diarthrodial joints are affected by rheumatoid arthritis. Erosive disease affects all of the joints in the hands and wrists, including the distal interphalangeal joints. Although these joints do not appear to be clinically involved, they have erosions in 10% of patients. Radionuclide scans in osteoarthritis and psoriatic arthritis demonstrate increased uptake in the distal interphalangeal joints similar to that found in Heberden's nodes.

The shoulder joint is commonly involved in rheumatoid arthritis. In addition to changes in the glenohumeral joint, erosive and cystic changes may be seen on the superolateral aspect of the humeral head adjacent to the greater tuberosity. Erosions may also be seen on the inferior border of the distal clavicle at the coracoclavicular ligament insertion. Rotator cuff atrophy or tear is common in severe disease. It can be recognized radiographically by progressive elevation of the humeral head in relation to the glenoid cavity and narrowing of the space between the top of the humerus and the under surface of the acromion. Mechanical erosion may also develop on the medial surgical neck of the humerus against the inferior glenoid, with elevation of

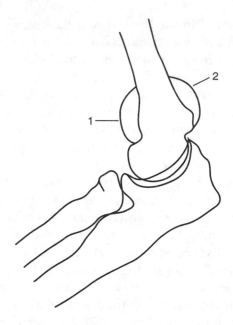

Figure 4-9. Fat pads on the lateral radiograph film of the elbow flexed at 90°. The anterior fat pad is normally seen in the coronoid fossa as a straight line. The posterior fat pad is not normally visible. The anterior fat pad becomes readily visible as shown (1), even with a small effusion. With large effusions the posterior fat pad becomes evident (2).

the humeral head. Bursal distension and Baker's cysts may occur in severe shoulder disease and can be visualized by arthrography.

Rheumatoid arthritis frequently involves the elbow joint and often results in severe limitation of movement. In early stages of the disease, an effusion may result in a positive "fat pad" sign (Fig. 4-9). An olecranon bursitis and subcutaneous nodules may be evident on a radiograph. Although subcutaneous nodules do not cause destruction of bone, they may lead to a smooth scalloped defect in the cortex of the ulna. Rheumatoid nodules never contain calcification, which helps to differentiate them from gouty tophi.

Hip disease is usually bilateral and loss of joint space is concentric, involving all of the femoral and acetabular surfaces. The femoral head migrates inward along the axis of the femoral neck (axial migration) and can ultimately protrude into the pelvis (protrusio acetabuli or Otto pelvis). Collapse of the femoral head may occur, especially in patients with severe disease on long-term corticosteroid therapy. Although this is often ascribed to avascular necrosis, in our opinion it is usually due to osseous collapse of osteoporotic bone. There is some evidence

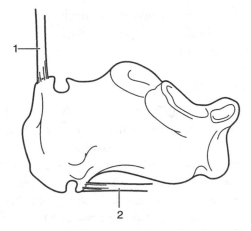

1—

2

Figure 4-10. Erosions of the calcaneum in rheumatoid arthritis.

that indomethacin may be associated with this complication, perhaps due to its effects on blood vessels, reducing the blood supply to the femoral head.

The three compartments of the knee joint are almost always involved in rheumatoid arthritis. The presence of joint effusion is usually the first abnormality seen on a radiograph, with opaqueness of the suprapatellar pouch and increased radiodensity in the posterior recesses. Erosions in the knee occur on the lateral and medial margins of the tibia and femur. Baker's cysts on the posterior aspect of the knee are best confirmed by ultrasonography. However, they can be better outlined by arthrography.

A number of other joints may also be affected by rheumatoid arthritis. All of the joints in the feet and ankles may become severely eroded in rheumatoid arthritis. Retrocalcaneal bursitis can be visualized on a radiograph as a soft tissue mass on the posterosuperior aspect of the calcaneus. The calcaneus may erode at the insertion of the Achilles tendon and on the undersurface of the calcaneus (Fig. 4-10). The manubriosternal joint can become severely eroded, leading to impaired respiratory function. In our experience, this is associated with severe cervical spine disease and kyphosis. The temporomandibular joint can become eroded, causing backward displacement of the mandible. The sacroiliac joints may be eroded in severe disease, but seldom cause symptoms.

Rib notching on the superior aspects of the second, third, fourth, and fifth ribs, which occurs a small distance from their angles, can result from an enthesopathy (e.g., inflammation at the insertion of the serratus posterior superior). Rib notches should not be confused with rib erosions in coarctation of the aorta, which occur on the inferior margin of the third to eighth ribs due to pressure from hypertrophied intercostal arteries. The upper third, fourth, and fifth ribs may be shaved away on their posterior aspect due to pressure from the vertebral border of the scapula. This type of erosion can lead to complete disappearance of a rib.

Juvenile Rheumatoid Arthritis

Juvenile rheumatoid arthritis differs from adult-onset rheumatoid arthritis in that articular erosions and joint space narrowing tend to be late manifestations. Although periostitis is more common in juvenile disease, it is seldom as florid as it is in psoriatic arthritis of Reiter's disease. Bony ankylosis is also common, perhaps due to the patient testing seronegative for IgM rheumatoid factor. Growth disturbances are common, especially when the onset of arthritis occurs in early life. Accelerated osseous growth of bone due to hyperemic stimulation of the epiphysis may lead to local gigantism. Alternatively, there may be premature closure of the epiphysis, resulting in shortening of bone. Growth disturbance may also cause ballooning of the epiphysis. Epiphyseal compression fractures may occur in the weight-bearing epiphyses in the lower limbs and hands, especially in patients who are osteoporotic. Although joint subluxation also occurs in juvenile rheumatoid arthritis, synovial cysts are rare. Sacroiliitis and spondylitis occur in juvenile rheumatoid arthritis, particularly in boys who are HLA-B27 positive. Changes in the spine are discussed later.

Ankylosing Spondylitis

Although ankylosing spondylitis is essentially a disease of the spine, peripheral joints are sometimes involved. The hip is perhaps the most common, and certainly the most disabling, joint involved. Similar to rheumatoid arthritis, joint space narrowing is more diffuse with axial migration of the femoral head. A feature of hip disease in ankylosing spondylitis is extensive osteophyte formation beginning on the lateral margin of the femoral head and progressing as a collar around the femoral neck. Bony ankylosis is common in ankylosing spondylitis. Other changes include loss of joint space, erosions, and juxta-articular osteoporosis. Due to restriction of motion and pain, hip arthroplasty is frequently required. New bone formation, often incorrectly referred to as ectopic calcification, is a common complication.

The shoulder joint is the second most commonly involved joint in ankylosing spondylitis. The entire outer aspect of the humeral head may be destroyed (the "hatchet" sign). As in rheumatoid arthritis, disruption of the rotator cuff may occur with similar clinical and radiologic consequences. Other joints involved in rheumatoid arthritis may also be involved in ankylosing spondylitis. However, erosive changes in ankylosing spondylitis are less common and bony ankylosis is more frequent, often occurring after a short period of time.

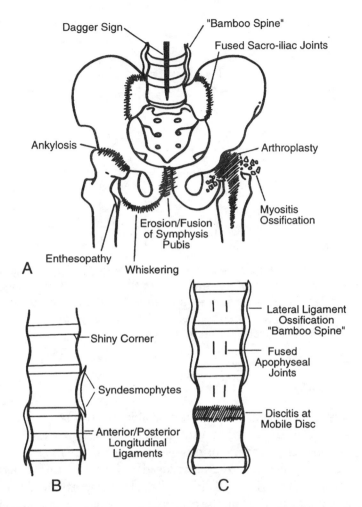

Figure 4-11. Ankylosing spondylitis. (A) anterior view, (B) lateral view, (C) antero-posterior view of spine. The radiologic findings that may occur in ankylosing spondylitis in the lumbar spine and pelvis.

A distinctive feature of ankylosing spondylitis is erosion and bone proliferation in cartilaginous joints, such as the symphysis pubis and manubriosternal articulation. In addition, enthesitis frequently occurs at tendon and ligament attachments to bone (e.g., the iliac crest, ischial tuberosity, femoral trochanter, humeral tuberosity, patella, calcaneus, malleoli, and distal clavicle). The appearance of "whiskering" on the ischial tuberosities is also characteristic of ankylosing spondylitis (Fig. 4-11).

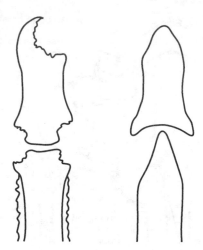

Figure 4-12. Typical changes in psoriatic arthritis. Note on the left the "mouse ear" appearance of the erosions in the distal interphalangeal joint and how cartilage (e.g., joint space) is spared. Note also the acro-osteolysis of the terminal phalanx and periostitis. On the right, a typical "pencil-cup" deformity. This can also occur in rheumatoid arthritis.

Psoriatic Arthritis

In psoriatic arthritis, changes in the peripheral joints may be identical to those in rheumatoid arthritis. However, other features differentiate the two diseases (Fig. 4-12). Although articular erosions occur in the marginal areas of the joints in rheumatoid arthritis, the cartilage, as judged by joint space on radiograph, appears to be spared in psoriatic arthritis until the later stages of the disease. These features result in the typical "mouse ear" appearance on radiographs. The distal interphalangeal joints are often eroded in the early stages of the disease. In severe disease, the joints may be whittled away, resulting in a typical "pencil cup" or "cup and saucer" appearance (Fig. 4-12). Arthritis mutilans may also develop.

A characteristic feature of psoriatic arthritis is destruction of the interphalangeal joint of the great toe. Resorption of the tufts of the terminal phalanges in the hands and feet is also characteristic of psoriatic arthritis. Osteolysis may progress to destruction of most of the distal phalanx. Periostitis is a common finding in psoriatic arthritis and is most often evident in the hands and feet. A phalanx can become increasingly dense due to new bone formation, resulting in an ivory appearance. Erosion of bone is often associated with irregular bony "paintbrush" excrescences. Changes in cartilaginous joints and at entheses are more pronounced in psoriatic arthritis than in ankylosing spondylitis. Although sacroiliitis occurs in 10%–25% of patients, association with HLA-B27 is less than in ankylosing spondylitis.

Reiter's Syndrome

The knees and ankles are invariably involved in acute Reiter's disease. Similar to psoriatic arthritis, bone proliferation is a prominent feature and a fluffy periostitis is common in the bones of the hands, feet, and in the area around the ankles. Enthesopathy is also a feature that is similar to psoriatic arthritis. Tendinous calcification and ossification may occur. Although sacroiliitis is usually considered unilateral, it is typically bilateral. If sacroiliitis is unilateral, Reiter's disease is likely. Similar to psoriatic arthritis, sacroiliitis is associated with HLA-B27, but not as frequently as in ankylosing spondylitis. Chronic Reiter's disease can be associated with destruction of multiple joints.

Enteropathic Arthropathies

An inflammatory polyarthritis complicates ulcerative colitis and Crohn's disease in approximately 25% of patients. Infrequently, joint space narrows and erosive changes develop. Although the arthritis in ulcerative colitis parallels the activity of the bowel disease, it does not in Crohn's disease. Sacroiliitis occurs in both conditions. The association with HLA-B27 is not as strong as it is in ankylosing spondylitis. Digital clubbing and bilateral symmetrical periostitis (e.g., hypertrophic osteoarthropathy) may develop in some patients with enteropathic arthropathy.

Gout and Other Crystal-Induced Diseases

In acute gout, radiographs usually only demonstrate soft tissue swelling. In chronic gouty arthritis, joint erosions produced by tophaceous deposits may also be seen. The erosions may be intraarticular, paraarticular, or present some distance from the joint. They often have a sclerotic margin giving a "punched-out" appearance and may have an overhanging margin (Fig. 4-13). The joint space is usually well preserved and osteoporosis does not occur. Bony ankylosis has rarely been reported. Gouty tophi in soft tissues produce asymmetrical and eccentric nodular opacities. Calcification may develop within tophi and, when present, helps to differentiate them from rheumatoid nodules.

Calcium pyrophosphate dihydrate crystal deposition disease (CPPD) has many causes. The most common cause is idiopathic (Table 4-6). Calcification may occur in hyaline fibrocartilage, synovium, and periarticular structures. What the first metatarsophalangeal joint is to gout, so the knee is to cal-

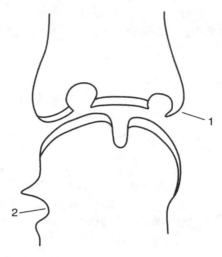

Figure 4-13. Joint and bone erosion due to gouty tophi. Note how there is preservation of cartilage and no loss of joint space despite intraarticular erosions. The marginal articular erosion shows a typical overhanging margin (1). Also note the erosion on the shaft of the bone at some distance from the joint (2).

cium pyrophosphate disease. Radiographs of the knees will show evidence of calcification in hyaline cartilage and menisci. Other common sites are the triangular fibrocartilage of the wrist and fibrocartilage of the symphysis pubis. In the knee, calcification tends to be more dense in the lateral meniscus. In the wrist, calcification can also be seen in hyaline cartilage in the carpal joints. Another feature of chondrocalcinosis in the wrist is a predilection for the trapezioscaphoid joint, which may only show bone sclerosis resembling osteoarthritis. In the absence of an erosive or degenerative arthritis, bone cysts in the radius and carpal bones are suggestive of chondrocalcinosis. In our ex-

Table 4–6. Causes of Calcium Pyrophosphate Crystal Deposition (CPPD) Disease

Idiopathic
Degenerative joint disease (controversial association)
Joint trauma (from hemorrhage)
Hyperparathyroidism
Hemochromatosis
Wilson's disease
Ochronosis
Hypothyroidism (doubtful)

Table 4–7. Conditions in which Calcium Hydroxyapatite Crystal Deposition Occurs

Collagen vascular disease (especially scleroderma and dermatomyositis)	Renal osteodystrophy
	Hypoparathyroidism
Dystrophic calcification secondary to inflammation and tumor	Hypervitaminosis D
	Milk alkali syndrome
Tumoral calcinosis	Myositis ossificans
Parasitic calcification	Enthesopathy (DISH)
Gouty tophus	CPPD crystal disposition

perience, septic arthritis of the wrist in elderly patients is usually due to an attack of acute pseudogout. Osteoarthritis of the hip with diffuse involvement of the joint is frequently due to chondrocalcinosis (Fig. 4-6). A severe destructive arthropathy can result from chondrocalcinosis.

Calcium hydroxyapatite crystal deposition disease characteristically occurs in the shoulder. Deposits occur in the supraspinatus tendon as thin-lined or rounded cotton wool-like deposits, which may become dense and well demarcated. Other conditions associated with periarticular calcification are listed in Table 4-7. Rupture of the supraspinatus aponeurosis, containing calcium hydroxyapatite crystals, can result in acute inflammation and resorption of the humeral head (Milwaukee shoulder syndrome).

Connective Tissue Diseases

A polyarthritis resembling rheumatoid arthritis may occur in systemic lupus erythematosus and progressive systemic sclerosis. Joint erosion and deformities are less common. In systemic lupus erythematosus, ulnar deviation of the metacarpophalangeal joints may occur without erosions (Jaccoud's arthritis). In progressive systemic sclerosis and dermatomyositis, calcification may be present in subcutaneous tissues and muscles. In progressive systemic sclerosis, calcification is a prominent feature with extensive subcutaneous, extraarticular, and occasionally intraarticular calcification. Acro-osteolysis of the tufts of the distal phalanges and severe resorption of the first carpometacarpal joint are also distinctive features of progressive systemic sclerosis. Punctate calcification may also be seen in the terminal phalanx. No other disease causes acro-osteolysis of the terminal phalanges and is associated with subcutaneous calcification. In dermatomyositis, subcutaneous and periarticular calcification are also present. The characteristic feature is sheet-like calcification present along muscle or fascial planes. Other causes of calcification and ossification in soft tissues are summarized in Table 4-8.

Table 4–8.   Calcification in Soft Tissues

Sites	Causes
Ligaments	Supraspinatus tendinitis
	Pellegrini-Stieda syndrome
	Seronegative spondyloarthropathies (at entheses, and ossification following joint surgery)
	Calcium pyrophosphate deposition disease
	Alkaptonuria (Ochronosis)
	Fluorosis
	Diabetes
Nerves	Leprosy
	Neurofibromatosis
Parasites and Infections	Cysticerci
	Guinea worm
	Loa Loa
	Tuberculosis
Trauma	Hematoma
	Burns
Connective Tissue Diseases	Scleroderma
	Dermatomyositis
	Ehlers Danlos syndrome
Tumors	Chondrosarcoma
	Hemangioma
	Liposarcoma
	Osteosarcoma
	Synovial osteochondromatosis
	Synovioma
Hypercalcemia	Hyperparathyroidism (especially secondary type)
	Sarcoidosis
	Hypervitaminosis D
	Milk alkali syndrome

Although there are many causes of avascular necrosis of bone, the rheumatologist encounters it most often in patients with systemic lupus erythematosus. In the early stages, plain radiographs and isotope bone scans are negative. The bone scans then become positive as a result of response of viable bone and hyperemia. Osteopenia becomes evident next and is followed by bone formation, causing increased density. The classic radiologic feature, the "crescent sign," is best seen in the hip joint in a frog-leg view (Fig. 4-14). Subchondral bone remains intact as it receives nourishment from overlying cartilage. Finally, the articular surface becomes flat. In avascular necrosis of bone cartilage, joint space remains normal.

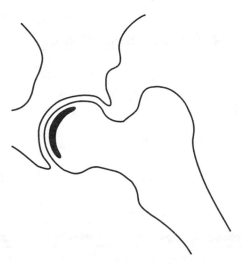

Figure 4-14. Typical "crescent sign" in avascular necrosis of the femoral head best visualized in the "frog" position. Note how there is no cartilage loss. This is a relatively late sign.

Hypermobility Syndromes

Joint laxity may cause joint pain, especially in young females. Arachnodactyly is characteristic of Marfan's syndrome, the most classic cause of joint laxity. Arachnodactyly can be confirmed by the metacarpal index, which is the sum of the ratios of the lengths of the second to fifth metacarpals to the width at the midpoint of their diaphyses. In healthy men the index is less than 8.8, while in healthy women it is less than 9.4. Arachnodactyly is also a feature of homocystinuria. Unlike Marfan's syndrome, homocystinuria is associated with multiple joint contractions and osteoporosis. The family doctor is not likely to encounter a patient with these disorders or other heritable disorders of connective tissue, such as Ehlers-Danlos syndrome and osteogenesis imperfecta. The family doctor is more likely to encounter a healthy patient with joint pains resulting from hypermobility. Although most family doctors and specialists record loss of joint range in clinical examination of the locomotor system, it is equally important to examine for increased joint range. In our experience, the hypermobility syndrome is a common and much overlooked entity. A useful and simple screening procedure is the nine-point Beighton scoring system (8). Patients have to perform the tasks shown in Table 4-9. Although the scoring is on an all or nothing basis, it is extremely useful (5). The radiologic features in "noneveryday" forms of arthritis and recommended texts are summarized in Table 4-10. The radiologic changes in systemic disorders are summarized in Table 4-11.

Table 4–9.   The Nine–Point Beighton Scoring System

	Right	Left
Dorsiflexion of the 5th MCP to 90 °	1	1
Apposition of thumb to volar aspect of forearm	1	1
Hyperextension of elbow by 10 °	1	1
Hyperextension of knee by 10 °	1	1
Hands flat on floor with knees extended	1	1
		1
Total		9

Table 4–10.   Radiologic Findings in "Noneveryday" Forms of Arthritis

Diseases	Radiologic features
Amyloid arthropathy	Bulky masses, especially around shoulders and wrists Joint space widening Well-marginated erosions Subchondral cysts Associated features of multiple myeloma
Rheumatic fever	Soft tissue swelling and effusions in acute stage Jaccoud's arthropathy (ulnar and flexion deformities of metacarpophalangeal joints and fibular and flexion deformities of metatarsophalangeal joints). No erosions
Hemochromatosis	Features of chondrocalcinosis with predilection for second and third metacarpophalangeal joints Osteoporosis
Wilson's disease	Findings of chondrocalcinosis, but distinctive irregular fragmentation of subchondral bone Osteoporosis
Ochronosis	Osteoporosis Disc degeneration with calcification Osteoarthritic changes in large peripheral joints, especially the hips
Pigmented villonodular synovitis	Erosive changes on both sides of joints Mild juxta-articular osteoporosis Evidence of effusion
Synovial chondromatosis	Multiple round calcified bodies Later changes of degenerative joint diseases
Hypertropic osteoarthropathy	Soft tissue swelling and occasional effusions No erosions Symmetrical periostitis

See references 6, 7, and 8 for related texts.

Table 4–11. Radiologic Changes in Bones and Joints in Systemic Disease

Diseases	Radiologic changes
Hyperparathyroidism	Subperiosteal resorption on radial side of digits, especially middle phalanx of index and middle fingers
	Resorption of tufts
	Osteopenia and fractures
	Endosteal and intracortical (tunneling resorption)
	Subchondral resorption with collapse in acromioclavicular and sacroiliac joints and symphysis pubis
	Brown tumors
	Patchy or diffuse bone sclerosis (especially in secondary hyperparathyroidism)
	Soft tissue calcification
	Chondrocalcinosis
Secondary Hyperparathyroidism	Similar to hyperparathyroidism, but periarticular and soft tissue calcification more marked with "tumoral" deposits
	Bone sclerosis more pronounced with sclerosis of vertebral endplates ("rugger-jersey" spine)
	Osteomalacia with Looser's zones (also due to aluminum toxicity in patients on dialysis)
	Beta-2 microglobulin amyloidosis with carpal tunnel syndrome, destructive spondyloarthropathy (especially in the cervical spine), and bone cysts
	Aseptic necrosis, especially in patients receiving corticosteroid therapy
	Septic arthritis and osteomyelitis due to infection from dialysis
Hypoparathyroidism	Localized or generalized osteosclerosis
	Subcutaneous calcification
	Premature closure of epiphyses
	Band-like increased density in metaphyses and vertebral endplates
	Osteoporosis
	Vertebral hyperostosis and enthesopathy
	Short metacarpals, metatarsals, and phalanges (especially first, fourth, and fifth in pseudohypoparathyroidism and pseudopseudohypoparathyroidism)
Hyperthyroidism	Osteoporosis
	Myopathy, especially around shoulder girdle
	Premature closure of epiphyses (in children)
	Clubbing of fingers, soft tissue swelling, and dense feathery periostitis of phalanges and metacarpals (thyroid acropathy) after treatment
Cretinism	Stippled epiphyses with fragmentation, especially in hips
	Delayed skeletal maturation and wormian bones
	Anterior beaking of vertebrae

Table 4–11.—continued

Diseases	Radiologic changes
Acromegaly	Thickening of soft tissues, especially noticeable in heel pad
	Widened joint spaces, bony outgrowths at tendon attachments, and spade–like tufts
	Enlarged bones with increase in height and breadth of vertebral bodies, which also show posterior scalloping
	Kyphosis
Scurvy (children)	Extensive subperiosteal new bone formation due to hemorrhage
	Osteoporosis (metaphyseal corner fractures)
	Sclerosis of epiphyseal rim (Wimberger's sign) and dense metaphyseal line (White line of Frankel); more proximal metaphyseal radiolucency
	Scorbutic rosary
Gaucher's disease	Erlenmeyer flask deformity at lower end of femur
	Avascular necrosis, especially in hip
	Osteoporosis with fractures
	Vertebral H-shaped endplate fractures
	Bone infarcts and neoplastic-like bone cysts
	Osteomyelitis not infrequent complication
Hemophilia	Hemorrhagic effusions
	Epiphyseal overgrowth and early fusion
	Secondary osteoarthritis changes, prominent subchondral bone cysts
	Widening of intercondylar notch in the knee and squaring of inferior pole of patella
	Enlargement of radial head and trochlear notch in elbow
	Pseudotumors of bone causing bone destruction
Sickle Cell Anemia	Dactylitis with periostitis and soft tissue swelling (hand-foot syndrome)
	Bone infarcts, especially in humeral and femoral heads, with H-shape in vertebral endplates
	Osteoporosis with coarsened trabeculae in widened tubular bones and widened diploic spaces in the skull
Thalassemia	Wide diploic spaces giving "hair-on-end" appearance
	Obliteration of paranasal sinuses
	Avascular necrosis of bone
Myelofibrosis	Sclerosis of bone marrow with increased density and cortical thickening of bones
Leukemia	Lucent band in metaphysis in children
	Periostitis
	Multiple scattered lucencies of variable size affecting cortical and cancellous bone ("moth-eaten" bone)
	Widening of sutures of skull in infants

Table 4–11.—*continued*

Diseases	Radiologic changes
Lymphoma	Lucent lesions giving rise when multiple to "moth-eaten" bone appearance
	Bone sclerosis (e.g., ivory vertebrae)
	Periostitis
	Bone fractures, especially in vertebrae
	Soft-tissue masses around bones (e.g., ribs, vertebrae)

Spine

Degenerative disease of the spine is the most common rheumatic disorder and, as demonstrated by paleopathologic studies, has been present since antiquity. Essentially, there are two processes involved. These include disc degeneration and osteoarthritis of the diarthrodial joints.

Degeneration of the intervertebral disc results in loss of disc space and sclerosis of bone in the adjoining vertebrae, similar to osteoarthritis in diarthrodial joints. The nucleus pulposus may herniate upward or downward into the vertebral body, forming Schmorl's nodes. They may also herniate posteriorly or posterolaterally, causing pressure effects on the cord or nerve roots. Osteophytes form as a result of traction by the anterior longitudinal ligament on its sites of attachment a few millimeters from the discovertebral junction (Fig. 4-15). Normally, the annulus fibrosus is firmly tethered to adjacent bone by fibers, the outer layer of which are known as Sharkey's fibers. When these fibers weaken, the annulus fibrosus moves forward, producing elevation of the anterior longitudinal ligament. As a result, bony overgrowth occurs where the ligament is attached to the vertebral body. The osteophytes are claw shaped or horizontal (traction spurs). The claw-shaped osteophytes may fuse ("kissing" osteophytes). Degeneration of an intervertebral disc can result in a vacuum phenomenon (Knutsson's sign), which may be centrally placed or located in the region of the annulus fibrosus. Herniation of a disc may be associated with a vacuum sign within the spinal canal. An intraosseous vacuum phenomenon usually indicates ischemic necrosis.

Osteoarthritis commonly occurs in the apophyseal joints and neurocentral or uncovertebral joints of Luschka. The radiologic changes are similar to those found in peripheral joints. Osteoarthritis may also develop in articulations formed as a result of congenital variations at the lumbosacral junction (e.g., between an enlarged transverse process of the fifth lumbar vertebra and

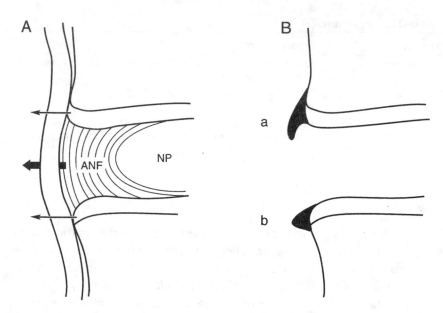

Figure 4-15. (A) anterior displacement of annulus fibrosus (ANF) causing traction on site of anterior longitudinal ligament attachment to the vertebral rim (NP = nucleus pulposus). (B) consequence of traction with formation of broad-based osteophytes of (a) claw and (b) traction types.

the wing of the sacrum or ischium). The neurocentral joints are only found in the cervical spine (C3–C7) and are best visualized in anteroposterior views (Fig. 4-16). Osteophytes from these joints can project into the intervertebral foramina, compromising nerve roots. They may also impinge on the vertebral artery. The costovertebral joints may also develop osteoarthritis, especially in the articulations of the lower two ribs.

Degenerative disc disease and associated osteoarthritis of the apophyseal joints occurs particularly in the lower cervical (especially C5–C6 and C6–C7) and lumbar spine (especially L4–L5 and L5–S1). Radiographs of these regions are of little value in evaluating neck and lower back pain, because abnormalities are found in healthy subjects as well as subjects with symptoms. Although spondylolisthesis is usually due to defects in the vertebral arch, it can be due to apophyseal osteoarthritis (Fig. 4-17). Apophyseal osteoarthritis is often seen at the L4–L5 level. Retrolisthesis may also rarely occur. When looking for spondylolisthesis, it is best to view the posterior border of the vertebrae rather than the anterior border. Spinal stenosis can also be diagnosed on conventional films (Fig. 4-18). Computed tomography is the preferred

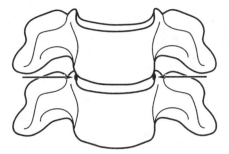

Figure 4-16. Anterior view of cervical vertebrae showing location of joints of Luschka.

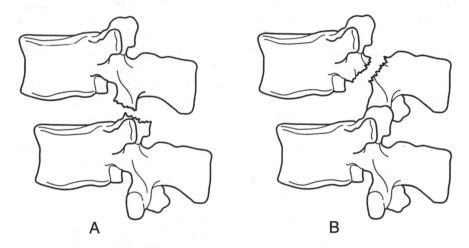

A   B

Figure 4-17. (A) spondylolisthesis without spondylolysis (e.g., without a defect in the pars interarticularis). This is pseudo-spondylolisthesis and the result of subluxation of the apophyseal joints. (B) spondylolisthesis with spondylolysis (e.g., with a defect in the pars interarticularis). Note that the step occurs below the level in pseudo-spondylolisthes and above the level in true spondylolisthesis. The former more frequently occurs between L4 and L5, and the latter between L5 and S1.

method because it can be used to visualize the lateral recesses (Fig. 4-19). Kyphosis in the elderly is commonly due to osteoporosis, especially in females (Dowager's hump). It may also be due to loss of height of the anterior part of the annulus fibrosus due to degenerative changes (Fig. 4-20).

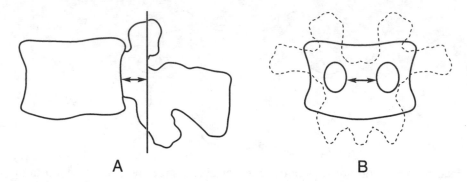

A   B

Figure 4-18.   Measurements to determine whether lumbar spinal stenosis exists. The anteroposterior diameter measured from the posterior surface of the vertebral body to a line drawn through the base of the superior process of the spinous process should not be lower than 15 mm (A). The transverse diameter should not exceed 20 mm and is measured by the distance between the inner aspects of the pedicles (B). Computed tomography provides the best diagnostic technique for assessment of lumbar spinal stenosis. An anteroposterior diameter less than 11.5 mm and an interpedicular distance less than 16 mm is generally regarded as indicative of lumbar stenosis. Even with a normal body, canal stenosis may occur with laminal thickening (see Fig. 4-19).

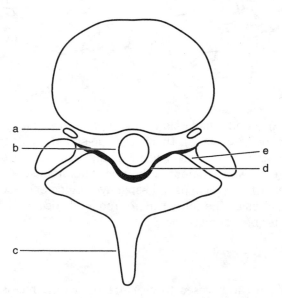

Figure 4-19.   What can be seen on computed tomography at the level of a lumbar disc. Note the emerging nerve routes in the lateral recesses (a). Other structures include thecal sac (b), spinous process (c), ligamenta flava (d), and facet joint (e).

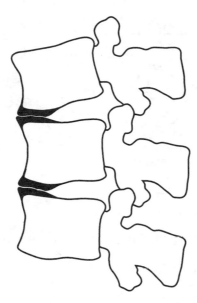

Figure 4-20. Senile kyphosis due to loss of height of anterior aspects of intervertebral discs and osteophyte formation. The anterior aspects of the discs may ankylose. In adolescents, irregular endplates and numerous Schmorl's nodes in the thoracic spine give rise to anterior disc space narrowing (Scheuermann's disease) with similar clinical consequences.

Radiographs of the thoracic and lumbar spines frequently reveal disc calcification. This usually occurs in the annulus fibrosus in elderly men and is due to deposits of calcium hydroxyapatite. Other causes of discal calcification are summarized in Table 4-12.

Table 4–12. Causes of Spinal Disc Calcification

Degenerative disc disease	Diffuse idiopathic skeletal hyperostosis (DISH)
Alkaptonuria (Ochronosis)	
Calcium pyrophosphate dihydrate deposition disease	Gout
	Idiopathic
Ankylosing spondylitis	Spine fusion
Juvenile rheumatoid arthritis	Natron*
Hemochromatosis (outer fibers of annulus)	

Natron is a natural sodium sesquicarbonate (Na_2CO, $3Na$, HCO_3, $2H_2O$) used in mummification. Hence, the reason why so many Egyptian mummies were thought to have ochronosis.
(Adapted with permission from: Weinberger A, Myers AR. Vertebral disc calcification in adults: a review. Semin Arthritis Rheum 1978; 18:69–75).

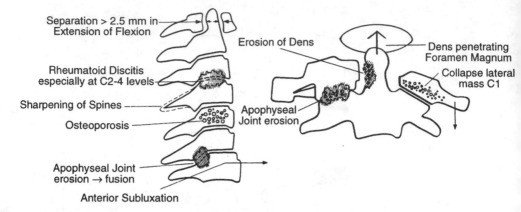

Separation > 2.5 mm in Extension of Flexion

Erosion of Dens

Dens penetrating Foramen Magnum

Rheumatoid Discitis especially at C2-4 levels

Collapse lateral mass C1

Sharpening of Spines

Apophyseal Joint erosion

Osteoporosis

Apophyseal Joint erosion → fusion

Anterior Subluxation

Figure 4-21. Rheumatoid arthritis of the cervical spine.

Rheumatoid Arthritis

The cervical spine is commonly affected in adult and juvenile rheumatoid arthritis. The changes that may occur in the cervical spine are summarized in Figure 4-21. Changes in the atlantoaxial articulations occur in at least 25% of patients with adult rheumatoid arthritis. Radiographs should include full flexion, extension, and lateral views, and an anteroposterior view through the open mouth. The distance between the posterior aspect of the anterior arch of the atlas and the anterior aspect of the odontoid process should not exceed 2.5 mm in full flexion when measured at the inferior aspect of the space. In addition to anterior displacement of the axis, vertical subluxation (cranial settling) may also occur. This can be defined by the distance of the tip of the odontoid peg above McGregor's line (drawn from the posterior margin of the hard palate to the most caudal part of the occipital curve). The distance should not be more than 4.5 mm. The anteroposterior radiograph view through the open mouth is useful in diagnosing vertical subluxation, in that it is invariably present when the bodies of the atlas collapse. Odontoid process erosion can also be identified. This usually occurs at the tip of the odontoid, but may cause complete disappearance of the dens. Periostitis and sclerosis of the dens are sometimes seen. The weakened and osteoporotic process may fracture. Lateral subluxation can also be diagnosed on anteroposterior radiographs. Inferior and posterior subluxations are rare. Atlantoaxial subluxation is often asymptomatic. Although it may cause pain, it seldom causes cord or nerve compression. However, vertical subluxation is extremely serious and often leads to death. Neurologic consequences only arise when anterior subluxation exceeds 9 mm.

Subaxial subluxation, which is usually anterior, is more likely to cause cord compression. Subluxations usually occur at the C3–C4 and C4–C5 levels. Mul-

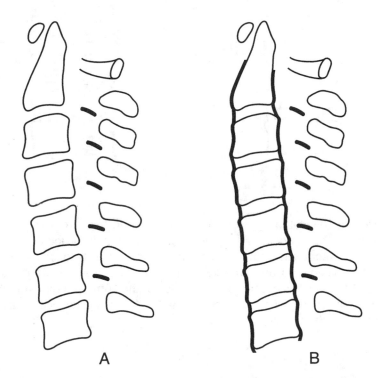

Figure 4-22. (A) Note how in juvenile rheumatoid arthritis there is fusion only of the apophyseal joints, while in ankylosing spondylitis (B) ossification also occurs in the anterior and posterior longitudinal ligaments. In juvenile rheumatoid arthritis, fusion of the apophyseal joints causes growth disturbances in the bodies (not shown in diagram).

tiple subluxations may produce a "doorstep" or "stepladder" appearance. Subluxations are primarily due to rheumatoid discitis, which many believe is the result of encroachment of inflamed synovium from the neurocentral joints into the disc spaces. Apophyseal joint erosions are commonly present. In severe cases, the spinous processes may be "sharpened" by inflammation in adjacent ligaments. The vertebral bodies are frequently osteoporotic.

Atlantoaxial subluxation is common in juvenile rheumatoid arthritis. Apophyseal joint involvement may lead to bony ankylosis. In juvenile rheumatoid arthritis, growth disturbances of the vertebral bodies are also common (Fig. 4-22).

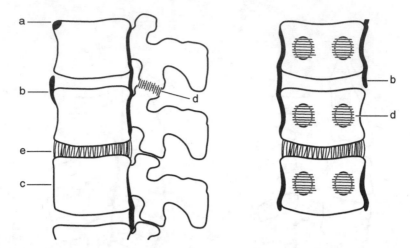

Figure 4-23. Typical changes of ankylosing spondylitis. Note the Romanus lesion (a); syndesmophytes (b); squaring of the vertebral bodies (c); fusion of the apophyseal joints (d); bamboo appearance, due to ossification of the longitudinal ligaments and preservation of the disc spaces, but destructive discitis at one level where there is no fusion of the corresponding apophyseal joint (e); this pseudoarthrosis occurs especially at the cervicothoracic and thoracolumbar junctions.

Ankylosing Spondylitis

In ankylosing spondylitis, early changes in the spine can be seen at the superior and inferior portions of the vertebral bodies at the anterior margins of the discovertebral junction (Fig. 4-23). They are termed "Romanus lesions" and are the result of osteitis at the attachments of the anterior longitudinal ligament. Initially, they appear as erosions. When they heal, they become radiodense, producing shiny corners. The lateral radiograph view shows squaring of the normal concavity of the anterior vertebral surface. Syndesmophytes are the result of ossification of the annulus fibrosus and appear as thin vertical outgrowths. Ossification also occurs in the anterior, lateral, and posterior ligaments (posterior ligaments result in the "dagger" sign), finally producing the typical "bamboo" spine (Fig. 4-11). Syndesmophytes have a broader base and rise higher up from the discovertebral junction than osteophytes. Fusion of the apophyseal joints may result in total lack of mobility. The ankylosed spine is particularly vulnerable to fracture. Destructive discitis (Andersson lesions) can occur, especially at the junction of the thoracic and lumbar spines, normally the most mobile part of the spine. Destructive discitis is often associated with a fracture through a neighboring ankylosed apophyseal joint. The radiologic appearance of destructive arthritis may resemble septic or neuropathic discitis. Calcification of intervertebral discs is common in patients with complete spinal fusion (Table 4-12). Compression of the cauda equina is particularly common in patients with spinal stenosis, which is best diagnosed by computerized tomography.

Psoriatic Arthritis
Spinal involvement in psoriatic arthritis is typically spotty (e.g., not uniform). The changes are frequently unilateral and asymmetrical. Romanus lesions and squaring of the anterior surfaces of the vertebrae are infrequent, although changes in the apophyseal joints are common. Thick, fluffy, bulky outgrowths of ossification are typical of psoriatic arthritis and unlike the changes in ankylosing spondylitis.

Reiter's Syndrome
The radiologic changes in the spine in Reiter's syndrome are similar to those seen in psoriatic arthritis with massive paravertebral ossification.

Enteropathic Arthropathies
In these diseases, the changes in the spine are identical to those in ankylosing spondylitis.

Diffuse Idiopathic Skeletal Hyperostosis
Diffuse idiopathic skeletal hyperostosis (DISH), also known as Forestier's disease and ankylosing hyperostosis, results in extensive changes in the soft tissues surrounding the vertebra bodies. Although there is no sacroiliitis, apophyseal joint ankylosis, or association with HLA-B27, it has often been mistaken for ankylosing spondylitis. Bulky flowing ossification of the anterior longitudinal ligament occurs, which has the appearance on radiograph of candle grease dripping down the spine (Fig. 4-24). The annulus fibrosus is also involved, but disc height is preserved and bone density is normal. It has been suggested that four contiguous vertebrae are involved, which would exclude spondylosis deformans as a possible diagnosis.

Lucent defects, or "windows," are one of the radiologic hallmarks of

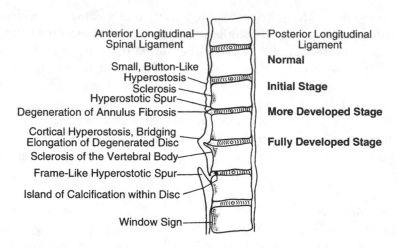

Anterior Longitudinal Spinal Ligament — Posterior Longitudinal Ligament

Normal

Small, Button-Like Hyperostosis — Sclerosis — Hyperostotic Spur —

Initial Stage

Degeneration of Annulus Fibrosis —

More Developed Stage

Cortical Hyperostosis, Bridging Elongation of Degenerated Disc — Sclerosis of the Vertebral Body — Frame-Like Hyperostotic Spur — Island of Calcification within Disc —

Fully Developed Stage

Window Sign —

Figure 4-24. Diffuse idiopathic skeletal hyperostosis.

DISH, appearing between the ossified ligament and vertebral bodies. The anterior aspect of the cervical spine is commonly involved, where DISH may lead to dysphagia. Occasionally, the posterior longitudinal ligament in the cervical spine becomes ossified, causing pressure on the cord. Ossification of the posterior longitudinal ligament in the cervical spine may be the sole manifestation of the disease. Changes in the thoracic spine are usually anterolateral, but only on the right side. The pulsating aorta on the left is believed to prevent changes on this side. DISH is also apparent in the lumbar spine, especially in the L1–L3 level. Although the upper nonarticular portions of the sacroiliac joints may be bridged by ligamentous ossification, erosive changes in joints are rarely seen. This is a purely productive, not erosive, condition. Enthesopathy is also common, especially around the insertions of the Achilles tendon and plantar aponeurosis, anterior surface of the patella at the quadriceps insertion, and the pelvis. Other than minor backache and stiffness, the condition is asymptomatic. It is present in elderly men, especially those with diabetes mellitus. The cause is unknown.

Miscellaneous Conditions
Many diseases can affect the spine, resulting in characteristic radiologic appearances. These are illustrated in Figures 4-25 and 4-26. The differentiation between septic discitis and intervertebral disc degeneration is summarized in Figure 4-27.

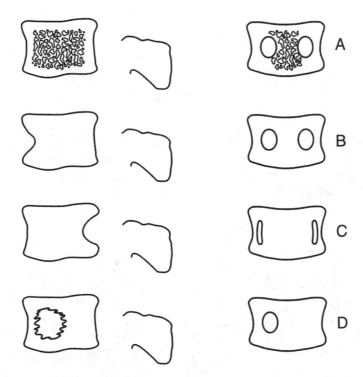

Figure 4-25. (A) hemangioma; note coarse trabecular pattern. (B) anterior scalloping of vertebral body due to pressure from aortic aneurysm, abscesses, and large tumor masses. (C) posterior scalloping of vertebral body; note widening of the interpedicular distance and flattening of pedicles in anteroposterior view. Often due to tumors in the spinal canal and occasionally seen in acromegaly. (D) metastatic carcinoma involves the body of the vertebra and the pedicle. The latter can cause disappearance of the pedicle giving the "winking owl" appearance. In osteoblastic metastasis the pedicle may become dense. The pedicle is not involved in multiple myeloma because it does not contain bone marrow.

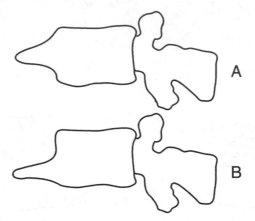

Figure 4-26. Anterior "beaking" of vertebral bodies. The central form (A) is unique to Morquio's syndrome, the lower form (B) to a number of conditions, including achondroplasia, cretinism, Down's syndrome, Hurler's syndrome, neuromuscular diseases, and pseudoachondroplasia.

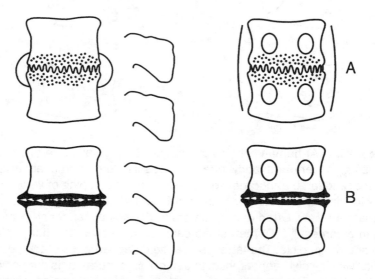

Figure 4-27. (A) septic discitis, which could be due to tuberculosis or pyogenic infection (e.g., intravenous drug users). Note soft tissue abscess formation, especially the bulging parasternal lines in the anteroposterior view. The stippled area represents osteoporosis. Also note absence of osteophytes. (B) intervertebral disc degeneration. Note the increase in bone density and formation of osteophytes.

Special Diagnostic Tests

Plain-film radiography is the most frequently used modality for evaluation of bone and joint disorders, especially traumatic conditions. However, many other imaging techniques are now available. Regardless of what new ancillary technique is employed, plain films should be available for comparison. The choice of imaging technique other than conventional films is determined by the type of suspected abnormality. For example, if plain films of the spine are negative in the patient with a breast tumor who is suspected of having metastases, a radioisotope bone scan should be done. If the bone scan is negative, magnetic resonance imaging (MRI) should be done. For a patient with systemic lupus erythematosus who is suspected of having avascular necrosis of bone, the preferred technique is MRI. New imaging techniques tend to be more expensive than conventional radiology. Although tomography is no more expensive than plain radiographs, an isotope bone scan and computed tomography (CT) are at least six times and the MRI over 12 times as costly.

Magnification Radiology

This is occasionally useful to enhance bony details, especially in the early detection of erosive changes in inflammatory joint disease. It is useful to use a hand magnifying glass when examining films for fine detail.

Tomography

Tomography is an important radiographic technique for evaluation of fractures, especially associated complications and their healing. With newly developed tomographic units, lesions as small as 1 mm can be detected. As a result, calcification within enchondromata or chondrosarcoma and a nidus of an osteoid osteoma can be identified. Although tomography can be useful in visualizing small erosions in inflammatory joint disease, it is seldom used for this purpose.

Computed Tomography (CT)

The advantage of CT over conventional radiography is that it can better demonstrate the anatomy of a slice of tissue and can provide images in the coronal, sagittal, and oblique planes. CT plays an important role in the evaluation of bone and soft tissue tumors. It can evaluate with great accuracy the extent of a bone lesion, whether there is a break in the cortex, and involvement of surrounding soft tissues. Iodinated contrast agents can be administered to enhance CT images. This has proved particularly valuable in evaluating disc disease, in which compression of the spinal cord or cauda equina is suspected. CT readily defines spinal stenosis and provides excellent visualization of the lateral recesses (Fig. 4-19). However, many radiologists now believe that MRI is better suited for investigation of disc prolapse, although this has not been confirmed. CT can be used to provide visible guidance for precise placement of a needle in aspiration of bone, spinal disc, or soft tissue lesion. Although CT is good at visualizing subtle abnormalities in bone, it is less satisfactory than MRI for soft tissue lesions. Also, in areas that contain metal (e.g., a joint prothesis) significant artifacts may be produced, degrading image quality.

Magnetic Resonance Imaging (MRI)

MRI is a complex, rapidly evolving diagnostic modality, in which images are obtained by passing radiofrequency pulses in the presence of a magnetic field. Alterations in the resonance of the nuclei of hydrogen atoms produce the image. Differences in the brightness of the images of tissues depend on the number of hydrogen atoms they contain and the rigidity of the tissues (e.g., the mobility of the hydrogen atoms). Cortical bone, which has virtually no mobile hydrogen atoms, will give a very low signal intensity (e.g., will appear black). Fibrocartilage, tendons, and ligaments also have low signal intensities. Fat, with highly mobile hydrogen atoms, has a high signal intensity. Hyaline cartilage also has a high signal intensity due to its high water content. Thus, it is possible to distinguish the nucleus pulposus from the annulus fibrosis. Muscle has a relatively low sign intensity, appearing grey.

The intensity of MRI images is also subject to technical variables chosen by the radiologist and to the T1 and T2 time constraints (e.g., the time it takes for the hydrogen nuclei to relax). By selecting different relaxation times, the radiologist is able to demonstrate the interface between two adjacent tissues having divergent relaxation times. Thus, the radiologist will be able to clearly differentiate between edematous tissue and skeletal muscle and between cerebrospinal fluid and spinal cord. The paramagnetic contrast agents [e.g., gadopentetate dimeglumine (Gd-DTPA) marketed as Magnevist by Berlex (Richmond, California)], administered intravenously, have proved extremely useful in enhancing MRI images. In reports, Gd-DTPA is often called gadolinium, which refers to the free ion that is extremely toxic. The metal chelate is metabolically inert and adverse reactions are relatively few and benign. Adverse reactions usually occur immediately after the injection and may include headache, nausea, vomiting, dizziness, paresthesia, and pain at the site of the injection and in other parts of the body (e.g., back, ear, eye, and teeth). Although hypotension can occur several hours after the injection, serious side effects, such as anaphylactic reactions and convulsions, are rare. Family practitioners should be aware of the adverse reactions to Magnevist when referring patients for MRI studies.

MRI provides direct coronal, sagittal, and multiplanar images as well as routine axial imaging. No untoward biologic effects have been reported. The procedure is noninvasive and uses no ionizing radiation. Serial examination can be safely performed. In addition to these advantages of MRI, the family doctor should also be acquainted with the disadvantages.

Apart from the cost, the major problem encountered with MRI is the dimension of the gantry. The small dimensions can induce claustrophobia, especially with the relatively long time the patient has to remain still during the examination. Anxiety can be relieved by mild sedation with oral diazepam (Valium). In young children who may not be able to cooperate, a general anesthetic may be required. Overweight patients [e.g., greater than 220 pounds (100 kg)] cannot be accommodated. Most radiologists prefer not to subject pregnant patients to investigation. Patients with electrical devices, such as cardiac pacemakers, metal implants, and nonmedical foreign bodies (e.g., shrapnel) should not be subjected to an MRI. However, if the shrapnel has been present for many years and is encased in fibrous tissue, it may be possible to perform an MRI. In these cases, the guidance of the radiologist should be sought. Most surgical clips are nonferromagnetic. Aneurysmal clips and cochlear implants are ferromagnetic and, consequently, are contraindications to MRI studies. Patients with intraocular lens implants, copper intrauterine devices, and joint protheses can safely undergo MRI investigations. MRI is dependent on the skill of the radiologist. One of the major limitations of MRI is limited availability of skilled radiologists and equipment.

MRI has proved superior to CT in the diagnosis of soft tissue abnormalities, including changes in bone marrow. Fine details of cortical bone and other calcified structures are better evaluated by plain radiographs or CT. MRI is the most sensitive imaging technique for detecting early changes in avascular necrosis of bone. It can detect changes before they can be detected by isotope bone scans. However, CT can more accurately define bone changes. On MRI, bone infarcts appear as a dark enclosed area surrounded by a bright region of fat ("ring" sign). The early diagnosis of osteomyelitis is best done by MRI. Conventional radiographs may show no changes for 10–14 days. An MRI will clearly define the extent of bone infection and can differentiate osteomyelitis from septic arthritis. CT scans are useful in demonstrating fissuring of bone, fat-fluid levels, and intraosseous gas. Radioisotope bone scans, using technetium-99, gallium-67, or indium-111 labelled leukocytes, are helpful in early diagnosis but are less accurate than MRI. It is often difficult to separate infection in the diabetic foot, especially osteomyelitis, ischemic changes due to vascular insufficiency, and neuropathic changes. Although indium-111 and gallium-67 labelled leukocyte scans are particularly useful in identifying infection, MRI has proved superior. CT demonstrates changes in bone, but is less able to differentiate osteomyelitis from neuropathic and vascular changes.

MRI is particularly useful in assessing tumors of bone and soft tissues. A smoothly contoured, well delineated, homogeneous mass is usually indicative of a benign tumor. An irregular, heterogeneous mass crossing tissue planes is suggestive of malignancy (Table 4-4). Metastases in bone can be identified by technetium-99 scans, which have the advantage of being less expensive. Before increased uptake of isotope is present, an MRI will often be positive.

MRI has limited application in peripheral joint disease. In the knee joint, meniscal and cruciate ligament tears, meniscal cysts, and osteochondritis can all be identified. MRI has recently proved useful in evaluating the surface of cartilage in knee patellar disorders, especially with gadopentetate dimeglumine enhancement. MRI often shows cartilage loss in osteoarthritis when no loss of joint space is apparent on conventional radiographs. Whether MRI will successfully assess cartilage loss in clinical therapeutic trials of antirheumatic drugs in osteoarthritis remains to be seen. MRI can also identify loose bodies in the knee joint when they are not calcified. Although Baker's cysts can be identified by MRI, this can be done more easily and cheaply by palpation, ultrasonography, and arthrography. Synovitis can be identified on MRI, but a diagnosis cannot be made, with the exception of pigmented villonodular synovitis. In this condition, the paramagnetic effects of hemosiderin result in focal areas of low signal intensity. MRI can identify thickened, inflamed plicae in the knee joints. These are embryonic remnants that persist into adult life in 20% of normal subjects. They may cause synovitis, cartilage damage, and a clicking sound as they pass over the femoral condyles.

In rheumatoid arthritis, CT and MRI are useful in identifying the effects of atlantoaxial and subaxial subluxations on the spinal cord. MRI has proved superior in identifying obliteration of the subarachnoid space, narrowing, and atrophy.

In disease of the spine, MRI has proved extremely useful in the diagnosis of extra- and intradural neoplasms, extra- and intramedullary neoplasms, congenital deformities, demyelinating diseases, and spinal cord injuries. Until recently, CT combined with myelography was the standard procedure for evaluating pressure effects on the spinal cord and cauda equina in lumbar disc prolapse. Recent studies suggest that MRI is superior for this evaluation. In septic discitis, MRI can identify changes earlier than CT. Although isotope bone scans are useful in localizing the diseased disc, they do not differentiate between infection and disc degeneration. Conventional radiographs have limited value in determining the presence of sacroiliitis, especially in young patients, because the sacroiliac joints are not isoplanar. Although CT scans have proved particularly useful in detecting sacroiliitis, recent evidence suggests that MRI may be more effective.

MRI has replaced arthrography in identifying rotator cuff tears of the shoulder. Arthrography cannot identify incomplete tears or tendinitis. However, CT scans are preferred when changes in the glenoid labrum are sought.

MRI is particularly useful in investigation of diffuse marrow disease. It identifies marrow infiltrates before changes are present on conventional radiographs and CT scans. Multiple myeloma is cold on bone isotope scans, but can be seen on MRI before bone changes occur.

MRI is a rapidly developing diagnostic modality. Much remains to be done to evaluate its value in comparison with other radiologic procedures in the diseases affecting the locomotor system. It is now possible to identify structures, such as nerves, and whether or not they may be compressed (carpal and tarsal tunnel syndromes). How valuable this and other uses of MRI will be in clinical practice remains to be determined. Regardless, MRI will be more extensively used as it becomes more available.

Ultrasonography

Ultrasonography is especially useful in evaluating tendons and bursae around joints and in determining whether soft tissue swellings are cystic or solid. It is not used for diagnosing bone abnormalities. Ultrasonography has many advantages, including cost. No ionizing radiation is involved and the procedure, which is noninvasive, can be performed at the bedside or in an operating room. Ultrasonography also allows comparison with the opposite side.

Ultrasonography can readily identify a Baker's cyst behind the knee and differentiate it from a popliteal aneurysm. Ruptures and tears of tendons around the shoulder joint can be demonstrated with ultrasonography. Calcium deposits in the supraspinatus aponeurosis are brightly echogenic on ultrasonography. Ultrasonography is also a useful adjunct to hip aspiration.

Arthrography, Myelography, and Discography

Arthrography involves the introduction of a contrast agent, which can be an iodide solution (positive contrast), air (negative contrast), or a combination of both. The iodide solutions currently used are nonionic and water-based [e.g., iohexol (Omnipaque)]. Arthrography is particularly useful in identifying intraarticular anatomical features. It is principally used to identify meniscal tears in the knee joint. However, tears of the lateral meniscus are often difficult to identify and are better done by MRI. Arthrography has also proved useful in demonstrating rotator cuff tears, adhesive capsulitis of the shoulder, and rupture of Baker's cysts behind the knee. Rupture of a Baker's cyst does not exclude a coexisting deep venous thrombosis. Contrast material can also be injected into tendon sheaths (tenontography) and bursae (bursography), but are seldom used. Other than transient pain and swelling, iohexol is seldom associated with adverse reactions when injected into joints. However, an allergic synovitis has been reported.

Myelography was originally performed using viscous iodized oil preparations. These often failed to fill nerve routes and led to an adhesive arachnoiditis if bleeding occurred during the procedure. Today, myelography is more effectively and safely performed using iohexol. The serious complication of adhesive arachnoiditis, although not entirely eliminated, is exceedingly rare. CT in conjunction with myelography is superior to either of these methods performed separately. Myelography with or without CT is gradually being replaced by MRI.

Patients referred for myelography must be questioned regarding previous reactions to contrast media, history of allergy (especially to iodine), and history of bronchial asthma. Although there are no absolute contraindications, special precautions are warranted. Patients receiving adrenergic beta-blockers may be more prone to severe reactions. Adrenaline must be cautiously administered for treatment of allergic reactions. Adverse reactions usually occur soon after administration of contrast media, are usually mild to moderate in severity, and seldom last long. Only a few fatalities have been reported with the use of iohexol for myelography. The most frequent reactions include headaches, nausea, vomiting, dizziness, and aseptic meningitis syndrome. Profound mental disturbances have been reported, but are rare. A thyrotoxic crisis or "storm" has been described following administration of iodinated contrast media in patients with thyrotoxicosis. MRI is preferred to myelography because it is noninvasive.

Discography is rarely performed today. The procedure has the advantage of being both a functional and radiologic study. During the injection of the nucleus pulposus, the patient is questioned regarding reproduction of symptoms. If symptoms are reproduced, the diagnosis of disc pathology is confirmed. The sensitivity of discography can be increased by combining it with a CT examination. The primary argument against discography is that it is an invasive procedure, which can lead to infection of the disc.

Scintigraphy

Three radionuclide preparations are used to perform bone scans. These are technetium-99m (99mTc) phosphate complexes, gallium-67 (67Ga), and indium-111 (111In) labelled white blood cells. The latter two are particularly useful in identifying sepsis. The uptake of 99mTc phosphate correlates with blood flow. Scintigraphy can image the entire skeleton. Although it has a high degree of sensitivity, the method lacks specificity. Nevertheless, scintigraphy has many uses in rheumatologic practice.

Osteolytic skeletal metastases only become apparent on conventional radiographs after 50% of bone has been lost. Scintigraphy identifies such metastases at an early stage and determines their extent and distribution. Osteosclerotic metastases are hot on radionuclide bone scans. In prostatic cancer, they are apparent before any changes are evident on routine radiographs or alterations are seen in alkaline and acid phosphatase enzymes. Primary malignant tumors of bone, such as osteosarcoma and Ewing's tumor, also appear hot, although Ewing's tumor may occasionally be cold. Nonossifying fibromas, solitary bone cysts, and bone islands are cold on isotope bone scans. Multiple myeloma is cold on scintigraphy, unless a fracture is present. Scintigraphy is most valuable in identifying an osteoid osteoma. This can be a small tumor, which is easily missed on conventional radiographs. A radionuclide bone scan classically demonstrates a double-density sign, with a high uptake in the nidus surrounded by a zone of low intake. A plain radiograph or CT will also demonstrate a radiolucent sphere surrounded by a circle of sclerotic bone.

Although scintigraphy is useful in the diagnosis of avascular necrosis of bone, it is less effective than MRI. The uptake of isotope is usually increased, but may decrease in the early stages due to ischemia. Scintigraphy usually demonstrates increased radionuclide uptake in osteochondrosis, such as Legg-Calvé-Perthes disease, Freiberg's osteochondritis of the metatarsal heads, Osgood-Schlatter's disease of the tibial epiphysis, and osteochondritis dissecans in the knee joint. However, there is an absence of radionuclide uptake in Kohler's osteochondritis of the tarsal navicular bone in the foot.

Scintigraphy is also useful in early diagnosis of osteomyelitis and septic arthritis. The use of 67Ga and 111In labelled white cells is particularly valuable. For example, when a 67Ga scan demonstrates a greater increase of uptake than that of 99mTc in bone, osteomyelitis is likely. A three phase radionuclide bone scan can often help to differentiate osteomyelitis from cellulitis. Osteomyelitis may show cold areas because of ischemia. Therefore, it is a wise precaution never to exclude osteomyelitis when scintigraphy is normal. Septic arthritis of the hip joint in children is a surgical emergency. Scintigraphy may demonstrate diminished or absent uptake in the femoral head due to ischemia. A 67Ga or 111In labelled white cell scan usually demonstrates increased uptake. It is better to aspirate or surgically drain the hip in a child clinically suspected of hip joint sepsis.

Scintigraphy often demonstrates stress fractures approximately 1–3 weeks before they are apparent on conventional radiographs. Although the method has a 100% sensitivity, it does not exclude other causes. Shin splints due to a diffuse periosteal reaction over the anterior portion of the tibia, which characteristically occurs in runners, can be confirmed by scintigraphy. When shin splints are present, scintigraphy demonstrates a diffuse periosteal uptake of radionuclide. Subtle fractures of the scaphoid, femoral neck, pelvis, and pars interarticularis in the lower lumbar spine are often demonstrated by scintigraphy when they are not apparent on conventional radiographs.

Metabolic disease of bone (e.g., osteomalacia), Paget's disease of bone, and hyperparathyroidism all show extensive uptake of radionuclides. In contrast, postmenopausal and senile osteoporosis show no increased uptake, unless fractures are present. The fractures appear as multifocal lesions.

In rheumatoid arthritis and other inflammatory joint diseases, increased uptake of radionuclide in inflamed joints may be present before clinical symptoms and signs of inflammation. Scintigraphy has proved of limited use in clinical therapeutic trials of antirheumatic drugs. Scintigraphy has not proved valuable in the diagnosis of sacroiliitis in ankylosing spondylitis or in other seronegative spondyloarthropathies. Diagnosis of sacroiliitis is best performed by CT and has recently been reported by MRI.

For those interested in further reading about these new diagnostic radiologic tests, a number of textbooks are now available (10—18).

Assessment of Osteoporosis

Osteoporosis is a prevalent disease without a curative treatment. In developed countries, it is predominant in elderly women. In the United States, it has been estimated to affect 15–20 million women over the age of 45, and is responsible for approximately 750,000 vertebral fractures and 250,000 femoral neck fractures. Femoral neck fractures are associated with significant mortality (10%–20% within 6 months) and postfracture morbidity, with approximately half of the survivors failing to gain social independence. Many risk factors have been identified, including race (whites and Asians), small stature and leanness (obesity protects due to loading affect on bone mass and conversion of androstenedione to estrogen precursors in fat cells), diminished calcium intake, excessive dietary protein and caffeine intake (resulting in increased urinary calcium loss), smoking, excessive alcohol intake, and decreased physical activity. However, none of these variables predict fractures in postmenopausal osteoporosis. Fractures can only be predicted by low bone mass, which has a statistical correlation with subsequent fractures. Other factors contribute to osteoporosis, such as a tendency of the elderly to fall.

The simplest and cheapest method of screening for osteoporosis is conventional radiography. The first evidence of osteoporosis is usually evident in the spine (Fig. 4-1). Singh and his colleagues have devised an index of severity of osteoporosis depending on the disappearance of trabeculae in the femoral head and neck (19,20). There are six grades:

1. Principal compressive trabeculae markedly reduced in number
2. Only principal trabeculae seen; all tensile trabeculae resorbed
3. Break in continuity of principal tensile trabeculae opposite greater trochanter
4. Principal tensile trabeculae reduced in number, but still traceable from lateral cortex to femoral neck
5. Principal tensile and compressive trabeculae accentuated and Ward's triangle prominent
6. All trabecular groups visible

Cortical thinning of bone can be measured by a hand lens and caliper. Although the second metacarpal bone is most frequently used for this purpose, other bones (e.g., humerus, clavicle, radius, femur, and mandible) have also been used. The basic measurements include the width of the cortex (A + B) and (C + D) and the diameter of the bone (E) at the midpoint of the bone, as illustrated for the second metacarpal in Figure 4-28. With these measurements, an index can be derived:

$$\frac{(A + B) + (C + D)}{E}$$

Although such indices have been widely used for population studies of osteoporosis, they have an error of 5%–10% and have been replaced by more sophisticated techniques. These techniques include single photon absorptiometry, dual photon absorptiometry, dual energy radiograph absorptiometry, and quantitative computed tomography. All of these methods have reasonable accuracy (reliability that the measured value reflects true mineral content), precision (longitudinal reproducibility in serial studies), and sensitivity (capacity to separate patients with osteoporosis from those with normal bones). In addition, these techniques have a reasonable radiation exposure and cost.

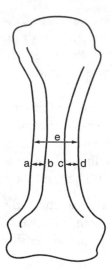

Figure 4-28. Measurement of cortical thickness at midpoint of second metacarpal. See text for formula for derivation of the index.

The single photon absorptiometry method essentially measures cortical bone in the radius and ulna, only 5% of which consists of cancellous bone. The method has also been applied to the os calcis, which provides a better assessment of cancellous bone because 80% of the os calcis is comprised of cancellous bone. Dual photon absorptiometry measures compact and cancellous bone in the lumbar spine (L1–L4) and neck and trochanter of the femur. The major disadvantage of this method is the inclusion of aortic calcification, osteophytes, posterior vertebral process, and the presence of compression fractures. Although dual energy radiograph absorptiometry can overcome these difficulties with lateral spinal views, this technique is not routinely performed. Dual energy radiograph absorptiometry is currently considered the most useful clinical method. In addition to measuring bone mass in the spine (L1–L4), it also provides measurement for the neck and trochanter of the femur, the radius and ulna, and the total skeleton. Quantitative computed tomography has also been used to provide a quantitative estimate of cancellous and cortical bone, both centrally and peripherally. However, it is more time-consuming than absorptiometry and has not become a routine procedure. Although ultrasonography has been used to assess the quantity and quality of bone mass in the patella and os calcis, it remains experimental.

There is considerable disagreement about the cost-effectiveness of initial and serial measurements of bone mass. However, bone density is the risk factor that has the highest correlation with subsequent bone fractures. Family doctors should consider bone density measurements in estrogen-deficient women (e.g., early menopause) and patients with evidence of osteoporosis in conventional radiographs. At least 30% of bone has to be lost before osteoporosis becomes evident in standard radiographs. Patients receiving corticosteroid therapy, long-term heparin therapy, or treatment for osteoporosis should have serial measurements of bone mineral density.

If a patient is found to be osteoporotic by one of these tests, treatment options must be considered. No single therapeutic agent inhibits bone resorption and stimulates bone formation, although there are a number that can bring about one of these results. Estrogen is an antiresorptive agent that appears to be effective in the early stages of menopause. Although estrogen slows the rate of bone loss, many clinicians are not convinced that it causes a clinically significant net gain in bone mass. There is evidence that estrogen therapy reduces the rate of fractures. The usual starting dose is 0.625 mg of conjugated equine estrogen (Premarin) taken daily by mouth or transdermally by patch twice weekly (Estraderm 50). If progesterone is given, there is less likelihood of endometrial hyperplasia and no risk of endometrial carcinoma. The major concern is whether prolonged estrogen therapy may lead to an increase in the rate of breast carcinoma. Most studies suggest that there is little, if any, risk during the first 10–15 years of treatment (21).

Biophosphonates are antiresorptive drugs that stabilize and increase bone mass and may even reduce vertebral fracture rate. Treatment is given orally [e.g., etidronate disodium (Didronel)], with 400 mg taken daily on an empty stomach at least 2 hours before a meal for 2 weeks. This is followed by a 10 week free period with a repeat of this 3 month cycle for 2 years. The absorption of etidronate disodium is obliterated when given concurrently with calcium. Therefore, patients should not ingest any calcium, either as tablets or in food, for at least 4 hours before and after taking the medication. The concurrent use of corticosteroid, phosphate, calcitonin, furosemide, or mithramycin therapies may result in addictive effects. Care must be taken in patients with renal failure because the drug is not biotransformed in the liver, but excreted unchanged in the urine. Mild diarrhea is the most common adverse effect.

Synthetic salmon calcitonin (e.g., Calcimar) also inhibits bone resorption and slows the rate of bone loss. The drug is given by subcutaneous injection, beginning with 25 IU three times a week, gradually increasing the dose over a 2 or 3 month period as required. Treatment is continued for as long as the drug is tolerated. Calcitonin has analgesic properties, probably due to the release of beta-endorphins, and is frequently used in patients who have sustained fractures.

Despite the fact that calcium supplements have been advocated for the treatment of osteoporosis, there is little evidence of an associated decrease in the rate of bone loss. However, it seems reasonable to have patients take 1.0–1.5 gm of calcium per day, especially those whose dietary histories suggest a low calcium intake. For adolescent girls, the daily dose recommended is 1.2 gm. During pregnancy, 1.6 gm/day is desirable.

Vitamin D has been recommended in the treatment of osteoporosis. It increases the calcium absorption and there is in vitro evidence that its active metabolites [e.g., 1,25 $(OH)_2D_3$ (calcitriol)] may stimulate bone formation. No more than 1000 IU/day of Vitamin D are required to assure adequate calcium absorption and 400 IU may be all that is needed. Whether the more potent metabolites, such as 0.25 μg/day of calcitriol, are required in the treatment of osteoporosis is debatable. It is prudent to keep a careful eye on elderly patients because they may overdose and develop hypercalcemia, with resulting effects on the heart or deterioration of renal function.

Sodium fluoride is widely used in oral doses of 50–70 mg/day in the treatment of osteoporosis. Although it stimulates bone formation, there is no evidence of reduction in fractures. Side effects include gastrointestinal distress, osteomalacia, fasciitis, and arthropathy.

Anabolic steroids are not currently recommended for treatment of osteoporosis. Treatment with parathyroid hormone is currently under investigation but is not yet recommended for routine use. Patients are encouraged to exercise to keep their bones strong and healthy, although it has not been proved that exercise causes such results. Patients treated with hydrochlorothiazide have sustained fewer osteoporotic hip fractures. Although thiazide diuretic therapy is not recommended for the treatment of osteoporosis, it may be useful to select thiazides as the diuretic of choice in patients with osteoporosis.

For those interested in further reading about the pathogenesis and treatment of osteoporosis, two excellent reviews have recently been published in The Lancet (22,23).

REFERENCES

1. Griffiths HJ. Basic bone radiology, 2nd ed. Norwalk, Connecticut: Appleton and Lange, 1987.
2. Brower AC. Arthritis in black and white. Philadelphia: WB Saunders Co., 1988.
3. Manaster BJ. Skeletal radiology. Chicago: Year Book Medical Publishers, Inc., 1989.
4 Forrester DM, Brown JC, Nesson JW. The radiology of joint disease. Philadelphia: WB Saunders Co., 1978.
5. Weston WJ, Palmer DG. Soft tissues of the extremities—a radiological study of rheumatic disease. New York: Springer-Verlag, 1978.

6. Madewell JE, Ragsdale BD, Sweet DE. Radiologic and pathologic analysis of solitary bone lesions. Part I: Internal margins. Radiol Clin N Am 1981;19:715-748.
7. Chapman S, Nakielny R. Aids to radiological differential diagnosis. London, England: Baillière Tindall, 1984.
8. Beighton P, Grahame R, Bird HA. Hypermobility of joints, 2nd ed. Berlin: Springer-Verlag, 1989.
9. Weinberger A, Myers AR. Vertebral disc calcification in adults: a review. Semin Arthritis Rheum 1978;18:69–75.
10. Resnick D, Niwayama G, eds. Diagnosis of bone and joint disorders. Vol 1–6. Philadelphia: WB Saunders Co., 1988.
11. Bassett LW, Gold RH, Seeger LL. MRI atlas of the musculoskeletal system. London, England: Martin Dunity Ltd., 1989.
12. Berquist TH, ed. MRI of the musculoskeletal system, 2nd ed. New York: Raven Press, 1990.
13. Enzmann DR, Delapaz RL, Rubin JB. Magnetic resonance of the spine. St Louis: CV Mosby Co., 1990.
14. Mink JH, Deutsch AL, eds. MRI of the musculoskeletal system—a teaching file. New York: Raven Press, 1990.
15. Crues JV III, ed. MRI of the musculoskeletal system. New York: Raven Press, 1991.
16. Markisz JA, ed. Musculoskeletal imaging: MRI, CT, nuclear medicine and ultrasound in clinical practice. Boston: Little, Brown and Co., 1991.
17. Quencer RM, ed. MRI of the spine. New York: Raven Press, 1991.
18. Stoller DW, ed. Magnetic resonance imaging in orthopaedics and sports medicine. Philadelphia: J.B. Lippincott Co., 1993.
19. Singh M, Macrath AR, Maini PS. Changes in trabecular pattern of the upper end of the femur as an index of osteoporosis. J Bone Jt Surg 1970;52A:457–467.
20. Singh M, Riggs BL, Beabout JW, et al. Femoral trabecular-pattern index for evaluation of spinal osteoporosis. Ann Intern Med 1972; 77:63–67.
21. Gorsky R, Koplan JP, Peterson HB, et al. Relative risks and benefits of long-term estrogen replacement therapy: a decision analysis. Obstet Gynecol 1994;83:161–166.
22. Dempster DW, Lindsay R. Pathogenesis of osteoporosis. Lancet 1993;341:797–801.
23. Lindsay R. Prevention and treatment of osteoporosis. Lancet 1993;341:801–805.

Laboratory Tests

5

Although laboratory tests are useful in supplementing clinical and radiologic findings, laboratory methods are only diagnostic in a few instances (e.g., isolation of bacteria from an infected arthritis; hypercalcaemia associated with hypophosphataemia and elevated serum alkaline phosphatase in hyperparathyroidism; and identification of sodium urate or calcium pyrophosphate crystals in acute gouty arthritis and acue chrondrocalcinosis). Laboratory tests are also important in arriving at a diagnosis and monitoring drug therapy.

SYNOVIAL FLUID ANALYSIS

A clinician should perform diagnostic joint aspiration to exclude hemorrhage, exclude or prove infection, and detect crystals (1,2). Arthrocentesis may be therapeutic (e.g., to relieve pressure in tense effusions and remove pus in septic arthritis) (3). Arthrocentesis should be performed according to the following procedures (4–7). The patient should be placed in a comfortable position and made to relax with reassurance and an explanation of the procedure. Shaving and drapes are not necessary. Although latex gloves were not worn in the past, they are necessary to protect the doctor from hepatitis and AIDS (8). The gloves do not have to be sterile. Talc and starch should not be used while putting on the gloves. The injection site can be marked with a ballpoint pen with the point withdrawn.

If the skin is cleaned with a detergent, such as povidone-iodine (Betadine), the patient should first be asked if they are allergic to iodine. However, allergies to iodine are unusual because the iodine is complexed with povidone (polyvinylpyrrolidine). After the skin is clean, chlorhexidine gluconate-cetrimide (Savlon) can be applied as a detergent and disinfectant. Many rheumatologists merely perform a quick application with an isopropyl alcohol (Duonalc Solution)-soaked swab. It is not known whether these procedures are necessary or less likely to introduce infection than simply injecting the needle through the skin, which is the procedure used by many rheumatologists. If a full sterile aseptic skin preparation is used, 1 minute of bacterial kill time following the application of the antiseptic is required to achieve a sterile field. Septic arthritis is a rare complication of intraarticular injection (probably less than 1/10,000) (8). Even if the skin cannot be sterilized and small pieces containing bacteria are introduced into the joint, the bacterial load is too small to cause infection (9–11). Joints should not be injected through skin that is abraided, diseased (e.g., psoriasis), or infected. Arthrocentesis may lead to septic arthritis if the patient is bacteremic.

If a fine bore needle (e.g., no. 25) is used, the injection site usually does not require a local anesthetic. If the patient is anxious or a large bore needle (e.g., no. 18) is used, a local anesthetic, such as 1% or 2% lidocaine hydrocarbonate (xylocaine), is desirable. When injecting the local anesthetic, the joint cavity should not be entered because this may interfere with the analysis. The skin can be anesthetized with an ethyl chloride spray, which has the added advantage of being sterile. The spray should be held steady and stopped at the first sign of freezing. The needle should be quickly thrust through the skin and then, to minimize hemorrhage, moved carefully through the synovial membrane into the joint cavity. If no fluid is obtained, the needle should be withdrawn back through the synovial membrane and redirected to penetrate the joint cavity in another place. Often, a needle is clearly in the joint space but no fluid can be aspirated. This is often due to blockage of the needle with a frond of synovium, piece of fibrin, or rice body. The needle can be cleared by injecting a small quantity of sterile saline.

Most useful information from synovial fluid analysis (e.g., total and differential white count, Gram's staining and cultures, and examination for crystals) can be performed with only 1–2 ml of fluid. A few drops may be all that is required for Gram's staining, cultures, and examination for crystals. Therefore, even a drop or two of fluid should not be discarded.

It is useful to have a small surgical forceps at hand while performing joint injections and aspirations because the needle may break during the procedure. The forceps can also be used to disconnect the needle from the syringe while switching syringes in the joint cavity.

Examination of Fluid

Gross Appearance

Normal fluid is clear and viscous, similar to an egg white (12). It is normally present in small quantities (e.g., 1 ml in an adult knee joint). If it is possible to read type through the aspirated fluid, the fluid is noninflammatory, with less than 500 cells/mm^3 (13). Normal fluid has approximately 100 mononuclear cells/mm^3. Patients with osteoarthritis and mechanical disorders (e.g., meniscal tears) usually have clear, noninflammatory fluids. The degree of opaqueness or turbidity reflects the total cell count. Fluids in which print can be read with difficulty (e.g., with slight turbidity) have cell counts between 500 and 5000 cells/mm^3. Such fluids are frequently seen in inflammatory arthritides, such as rheumatoid arthritis, seronegative spondyloarthropathies, and gouty arthritis. Fluids in which print can be seen but not read (e.g., moderate turbidity) can be expected to have total white cell counts between 5000 and 50,000 cells/mm^3. Such fluids are frequently seen in rheumatoid arthritis, seronegative spondyloarthropathies, gouty arthritis, and bacterial infection of joints. Fluids that are totally opaque (e.g., markedly turbid) can be expected to have cell counts exceeding 50,000 cells/cm^3. Although such fluids are especially seen in septic arthritis, they can also be seen in rheumatoid and gouty arthritis.

Normal and noninflammatory fluids retain their viscosity and drip from the end of a needle like syrup (string sign). Inflammatory fluids, except those that are grossly purulent, tend to drip like water.

The clots of mucin are poorly formed in inflammatory fluids. Mucin clots consist of protein hyaluronate formed by the additon of 0.2% glacial acetic acid (1 part of supernatant of centrifuged fluid to 4 parts of acid) and can be dispersed with hyaluronidase. The mucin clot test is obsolescent, if not obsolete.

Hemorrhage fluid may result from rupture of a blood vessel during arthrocentesis. In this case, the blood is not evenly dispersed in the fluid and easily clots. Uniformly red fluid indicates recent bleeding and may be so severe as to make it difficult to distinguish from pure blood. Hemarthrosis has many causes, trauma and coagulation disorders being the most common (Table 5-1). Repeated hemorrhage into a joint in the absence of a coagulation defect suggests pigmented nodular synovitis.

The presence of large fat droplets in synovial fluid suggests a fracture involving the marrow cavity. Fluids may be "milky" due to the presence of small fat droplets and are occasionally seen in rheumatoid arthritis and chronic gouty arthritis. In pancreatic-fat necrosis and fractures, small lipid droplets may be present in the synovial fluid aspirate and cause Coulter counters to be misread.

Table 5–1. Causes of Hemarthrosis

Common
  Trauma
    Arthrocentesis; blood unevenly mixed with synovial fluid. With or without fracture; blood usually evenly mixed
  Coagulation Disorders
    Therapeutic anticoagulants
    Hemophilia and Christmas disease
    von Willebrand's disease
    Other bleeding disorders
    Thrombocytopenia
    Essential thrombocytosis
  Idiopathic
Uncommon
    Hemangioma and anteriovenous malformation
    Pigmented villonodular synovitis; often recurrent episodes of bleeding
    Tumors; synoviomas and metastatic
    Ehlers Danlos syndrome and pseudoxanthoma elasticum
    Sickle cell disease
    Scurvy*

*This was common among Glasgow bachelors and widowers who had to cook for themselves. Fruits were absent from the diet. The only source of Vitamin C was the potato, which men found tiresome to peel. In North America chronic alcoholics are most likely to develop scurvy.

Cell Counts

Fluid for total and differential white cell counts should be put in green-topped heparin tubes (50 units/ml). The white cell count is best performed using saline (0.9 g/dl) as opposed to acetic acid, which may form mucin clots. Red cells can be lysed with 0.3 g/dl saline, which can be used instead of normal saline. It is useful to add methylene blue (0.1%) to differentiate white blood cells from other cells and particles. The white cell count is best performed using a white cell pipette and magnificaiton of 400. If only a few cells are present, the fluid should be centrifuged at 3000 rpm for 10 minutes and the sediment resuspended. Adding hyaluronidase facilitates identification of different cells (14). Due to the viscous nature of synovial fluid, it is not surprising that divergent cell counts are common (15,16).

In gonococcal and tuberculous arthritis, the white cell counts in synovial fluid may not be particularly high. In immunocompromised patients with septic arthritis, counts of less than 20,000 cells/mm^3 are common (15,17). Eosinophils may be prominent in differential cell counts in patients with parasitic infections; urticaria and dermatographism; Lyme disease; and, occasionally, rheumatoid arthritis (18–21). Monocytes may predominate in early

Table 5–2. Gram Stain Technique

1. Heat fix smear on clean slide
2. Treat with basic dye, crystal violet, for 1 minute
3. Wash with water
4. Add Gram's iodine solution* for 1 minute
5. Thoroughly rinse with water
6. Flood slide with decolorizer, acetone-alcohol, until no violet color washes off slide; this takes 10 seconds or less
7. Wash again thoroughly with water
8. Counterstain with red dye, safranin, for 1 minute
9. Rinse again with water, blot dry, and examine under oil immersion

*3% iodine; potassium iodide in water or weak buffer pH 8.0 to neutralize acidity formed from iodine on standing.

rheumatoid arthritis. In chronic rheumatoid arthritis, the phagolysosomes of the polymorphonuclear leukocytes (RA cells or rhagocytes) may be prominent due to ingestion of immunoglobulin G (IgG), immunoglobulin M (IgM) rheumatoid factor, and complement. Macrophages that have ingested polymorphonuclear leucocytes were first found in Reiter's disease, but are no longer considered specific to this condition (23). In systemic lupus erythematosus, LE cells and tart cells may be found in synovial fluid. Other cells that may be present include platelets, mast cells, synoviocytes, marrow cells (if fractures are present), Gaucher's cells, sickle erythrocytes, and malignant cells, especialy leukemic.

Bacteriology

All synovial aspirates should be cultured. A Gram stain may help to identify the organism and allow early institution of the appropriate antibiotic therapy. A useful procedure is to stain the synovial aspirates with methylene blue, which identifies bacterial morphology and cell type. A Gram stain (Table 5-2) should then be done to more accurately identify the organism and allow early institution of the appropriate antibiotic therapy. A Ziehl-Neelsen stain (Table 5-3) should be done if tuberculosis is suspected. In tuberculous arthritis, only one in five samples of synovial fluids are positive on Ziehl-Neelsen staining and, at most, only four in five are positive on culture. Tuberculin skin tests are usually positive in tuberculous arthritis. Only 50% of patients have evidence of tuberculosis on chest radiographs (23). Although tuberculous arthritis is usually chronic, it occasionally may be acute (23–26). Immunocompromised patients and patients who have received repeated intraarticular injections of corticosteroids may develop atypical mycobacterial

Table 5–3. Ziehl-Neelson Technique

1. Flood slide with carbol fuchsin and heat until steam rises; DO NOT BOIL
2. After 3–4 minutes, apply more heat until steam rises again; do not let the stain dry on the slide
3. 5–7 minutes after first application of heat, wash the slide thoroughly under running water
4. Decolorize with 3% acid alcohol until all traces of red have disappeared
5. Wash well in water
6. Counterstain with 0.5% malachite green (20–30 sec)
7. Wash and stand on end to drain; DO NOT BLOT

joint infections (27). Although fungal arthritis is usually indolent, it occasionally may be acute when infected with blastomycosis or candida (28). Cultures for fungi should perhaps be more routinely performed.

Gas liquid chromatography has not proved effective in detection of bacterial infection in joints (29). In the fasting state, the concentration of glucose (not total reducing substance) in synovial fluid is the same as it is in blood. In septic and tuberculous arthritis, synovial glucose concentration may be approximately 40 mg/dl (2.2 mmol/L) less than it is in blood. However, low serum glucose concentrations may also be decreased in noninfected inflammatory conditions (30). Although high levels of lactate in synovial fluid indicate bacterial infection, these levels are not used as a routine test (31).

Crystals

A wet smear of aspirated synovial fluid should be made as quickly as possible (4). Only one drop of fluid is required to identify crystals and should never be discarded as insufficient. Circumstances will dictate whether the drop should be sent for culture or examined for crystals. If examined for crystals, the slide and coverslip should be scrupulously clean and dust free, because many things that constitute dirt on a slide or coverslip are birefringent (e.g., mineral dust, paper tissue fibers) and may lead to difficulty in interpretation. A permanent preparation can be made by sealing the edges of the coverslip with nail varnish. Any crystalline material forming at the junction of the varnish and fluid should be ignored because nail varnish forms strongly positive birefringent rod-shaped crystals.

If the fluid is collected in a tube, sodium heparin should be the anticoagulant used because both lithium and oxalate form crystals that may be phagocytosed by polymorphonuclear leukocytes and can easily be confused with calcium pyrophosphate dihydrate crystals. There is usually no need to cen-

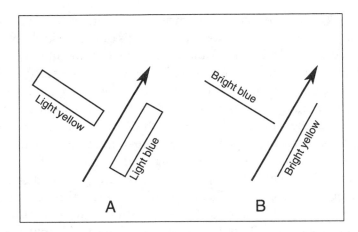

Figure 5-1.   Color of pyrophosphate (A) and urate crystals (B) under polarized light.

trifuge synovial fluid. Fluid should not be refrigerated because there is evidence that urate crystals may form in patients who are hyperuricemic. Refrigeration at 4°C does not slow the disappearance of crystals, with the exception of urate crystals, but does decrease the appearance of artifactual crystals.

In many laboratories, synovial fluid is kept overnight before being examined. A significant fall in white cell count can be expected, so that a mildly inflamed fluid may have a normal count after 6 hours. There is no difficulty in identifying urate crystals, which may be detected even after 2 months, although they are less numerous, less birefringent, and usually smaller than when the fluid was first aspirated. Crystals of calcium pyrophosphate dihydrate are more difficult to identify after 24 hours and, after 2 weeks to 2 months, will be almost completely gone. Storage of synovial fluid leads to the development of artifactual crystals. Therefore, it is important to examine synovial fluid as promptly as possible.

In acute gout, urate crystals can often be easily identified on a wet film by ordinary light microscopy (32). Other crystals, such as calcium pyrophosphate, are more difficult to identify. Therefore, the slide should be examined with a polarizing microscope (33). Urate crystals appear as long, sharp-ended aciculae. In acute gout, they are often present within polymorphonuclear leukocytes. Urate crystals are strongly negatively birefringent, in contrast to calcium pyrophosphate dihydrate crystals, which are positively birefringent (Fig. 5-1). Although urate crystals are necessary to initiate inflammation, they can be found in 20% of effusions in hyperuricemic patients who have never had joint inflammation (33,34). Acute gouty arthritis can occur in the absence of polymorphonuclear leukocytes. Urate crystals may be absent in sympathetic effusions associated with bursal gout and may be found several months after an acute attack of gout has subsided and the patient is asymptomatic (35). Treatment to reduce serum urate concentrations after an acute gout attack reduces the numbers of crystals found in the joint effusions. Urate crystals may appear as "beachball" spherules made up of numerous acicular crystals radiating out from the center and retaining their negative birefringence (36).

Recently, there have been reports of uric acid crystals and calcium pyrophosphate dihydrate crystals being found in septic arthritis. This appears to result from the septic process, which "strip-mines" the crystals from cartilage and synovium (7). Therefore, fluid should always be cultured, even if the clinical diagnosis is acute gout.

Although calcium pyrophosphate dihydrate crystals are typically rhomboidal in shape, they may also be acicular. They are smaller than urate crystals and positively birefringent (Fig. 5-1). Like uric acid crystals, calcium pyrophosphate dihydrate crystals are frequently intracellular (37,38).

Basic calcium phosphate crystals (apatites) include carbonate substituted hydroxyapatite, octacalcium phosphate, and, less commonly, tricalcium phosphate. They are classically present in calcific periarthritis of the shoulder and the Milwaukee shoulder syndrome (also known as idiopathic destructive arthritis of the shoulder, cuff tear arthropathy, and apatite-associated destructive arthritis), in which they are present in abundance in joint fluid associated with a low synovial fluid leukocyte count (less than 1000/mm^3).

Basic calcium phosphate crystals are too small to be seen by ordinary light microscopy, but can be identified with electron microscopy. Basic calcium phosphate crystals often clump to form microaggregates. With phase contrast microscopy, these microaggregates appear as "shiny coins" both within and outside of polymorphonuclear leukocytes. These shiny inclusions stain purple with Wright's stain and bright red with Alizarin red S (39). Alizarin red S is nonspecific because any calcium containing substance will stain the same color. Therefore, identification of basic calcium phosphate crystals is not easy. C^{14} diphosphonate probes have proved useful, but only as a research tool. Basic calcium phosphate crystals are frequently present in synovial fluid obtained from osteoarthritic joints, but appear to be related to severity of radiologic changes rather than indices of inflammation. In osteoarthritis, the crystals appear to originate from articular cartilage, along with other basic phosphates.

In rheumatoid arthritis, cholesterol crystals are occasionally found in synovial fluid. They are easily identified, appearing as large, flat rhomboid plates with one or more notched corners, often stacked on top of each other. Their clinical significance is unknown, although some rheumatologists believe they may cause mild inflammation. Cholesterol crystals are strongly birefringent and the occasional needle-shaped form, which is strongly negatively birefringent, may be mistaken for a uric acid crystal.

Liquid lipid crystals of phospholipids appear as tiny bubbles under light microscopy and have a Maltese cross appearance with positive birefringence in polarized light (40). They may be found within cells and stain with Sudan black B. Triglycerides are similar to phospholipids under light microscopy, but are not birefringent.

Oxalate crystals may be found in synovial fluid in patients with the rare, recessively-inherited primary oxalosis or oxalate accumulation secondary to long-term dialysis for renal failure and inflammatory bowel disease (41). Ascorbic acid is biotransformed to oxalate and, when given in large amounts to patients receiving dialysis, may aggravate oxalate deposition. Oxalate crystals appear in two forms; the dihydrate is bipyramidal and the monohydrate is polymorphic. Some crystals have strong birefringence while others have weak or no birefringence. They may be intracellular and can easily be confused with calcium hydroxyapatite. The crystals stain bright red with Alizarin red S due to the presence of calcium. Oxalate used as an anticoagulant forms calcium oxalate crystals.

Preparations of corticosteroids used for intraarticular injection are crystalline. As a result, they may cause a gout-like post-injection flare, which cures itself when the crystals dissolve. However, these crystals may still be found after a month. The crystals are different sizes and shapes, brightly birefringent (mostly positive), and frequently intracellular.

Many other crystals and refractile and nonrefractile artifacts can be found in synovial fluid. Immunoglobulins and cryoglobulins can crystallize in the joints of patients with multiple myeloma and cryoglobulinemia. They appear as positively birefringent rhomboids or rods. In amyloid arthropathy, complicating multiple myeloma, and the β2 microglobulin amyloidosis associated with chronic renal dialysis, amyloid deposits may appear as large doughy masses. Amyloid deposits can be easily identified by their typical apple-green birefringence with Congo red staining. Cartilage shards, which normally appear as colorless, birefringent chunks, appear blue-black in ochronosis. Cartilage shards are unlikely to be confused with crystals, especially if chondrocytes are seen. Fibrin filaments are nonbirefringent and, if straight, might be confused on light microscopy with uric acid crystals. Charcot-Leyden crystals occur in eosinophilic synovitis and appear as bipyramidal hexagonal crystals with weakly positive, negative, or no birefringence. Hemoglobin and hematoidin crystals may be found in patients with hemarthrosis. Hemoglobin crystals are rectangular and often found within erythrocytes. They are weakly positive or negatively birefringent. Hematoidin crystals are intensely birefringent, with positive and negative elongation. Under ordinary light microscopy, hematoidin crystals have a golden-brown color.

Although thorns in synovial fluid are birefringent, they are usually easily identified. Computed tomography (CT) and magnetic resonance imaging (MRI) often prove useful in identifying foreign bodies when plain radiographs are unrevealing. Detritic synovitis is due to worn-down silastic fragments, which are glossy and irregular in shape and size, but not birefringent. Polymethylmethacrylate and high molecular weight polyethylene debris from joint protheses are birefringent, but unlikely to be confused with urate or calcium pyrophosphate dihydrate crystals. Talc (magnesium silicate) and starch from surgical gloves may appear as bright Maltese crosses.

Biochemical and Immunological Tests

Synovial fluid is a dialysate of plasma with a secretion of hyaluronic acid from Type B synovial cells. Although synovial fluid protein concentrations are less than that of plasma, they increase in inflammatory joint disease. Estimation of synovial proteins is only useful if studies on complement are being performed (31). In rheumatoid arthritis, both the total 1-(CH_{50}), and C_3 and C_4 components are reduced due to formation of immune complexes. However, their estimation is of little practical value. Measurement of IgM rheumatoid factor and antinuclear autoantibodies is also unvaluable because their titer is the same as that in serum. Antibiotics reach high concentrations in synovial fluid after oral and parenteral administration and do not require assay during treatment of septic arthritis. The measurement of antirheumatic drug concentrations in synovial fluid is not routinely performed in clinical practice (42).

Dr. Joseph L. Hollander and his colleagues first coined the term "synovianalysis" to emphasize the importance of examination of synovial fluid in joint disease analogous to urinalysis in disease of the kidney. Although in clinical practice the most important crystals to recognize are uric acid and calcium pyrophosphate dihydrate crystals, doctors should be aware that many things might glow in synovial fluid (e.g., when examined by a polarizing microscope). The reader is referred to the two monographs on synovial fluid analysis by Dr. Schumacher and his colleagues, and those by Doherty et al, and Steinbrocker and Neustadt (4,5,37,43).

BIOPSIES

Synovial Biopsy

Although synovial biopsy is rarely required to establish a diagnosis of joint disease, it is mandatory in unexplained monoarthritis. The three available methods of biopsy are needle biopsy, arthroscopic biopsy, and biopsy by arthrotomy. Biopsy by arthrotomy is the most popular method with pathologists because it provides a more generous specimen. Open biopsy is necessary for fibrocartilaginous joints, such as the sacroiliac joints, and small joints, such as those in the fingers and toes.

Today, many rheumatologists prefer a needle biopsy using one of the many needles that are available (Parker-Pearson, Polley-Bickel, Williamson-Holt, Vim-Silverman, and Cope). The Parker-Pearson needle is the most popular and the method has been well described by Drs. Schumacher and Kulka (44). The major problem with needle biopsy is the relatively small sample of tissue obtained.

Arthroscopy is popular with orthopaedic surgeons and, similar to open biopsy, has the advantage of direct visualization and optimal site selection for biopsy. It also allows identification of discrete localized lesions. Although arthroscopy was performed only on large joints, such as the knee, more sophisticated technical methods have made an increasing number of smaller joints accessible to the procedure. Arthroscopy can be performed in an ambulatory surgical unit using a local, regional, spinal, or general anesthetic. Severe contraction of a joint, hemorrhagic diathesis, and generalized sepsis are contraindications. The procedure has been well described (45–47).

Synovial tissue, once obtained, should be fixed in neutral buffered formalin, unless gout is suspected. In this case, absolute alcohol should be used to prevent the dissolution of crystals. It is prudent to always send biopsy specimens for culture. Culture should include ordinary organisms, acid-fast bacilli, and fungi. A histologic diagnosis can be made with confidence in conditions such as pigmented villonodular synovitis, synovial chondromatosis, sarcoidosis, septic arthritis, ochronosis (fragments or chards of pigmented cartilage), amyloidosis (congo red positive), synovioma and metastatic cancer, Whipple's disease [periodic acid-Schiff stain (PAS) positive macrophages], and multicentric reticulohistocytosis (histiocytes and multinucleated giant cells) (24). None of these conditions are "everyday" forms of arthritis and are rare. Histology is not a "gold standard" for these conditions. For example, the changes in sarcoidosis may not make it possible to exclude tuberculosis and may be entirely nonspecific (48).

Although synovial biopsy is claimed to have prognostic value in rheumatoid arthritis, it is not helpful in making a diagnosis of the common forms of arthritis, such as rheumatoid arthritis and the seronegative spondyloarthropathies (49). In these diseases, identical changes may be found. In rheumatoid arthritis and other iflammatory arthritides, there is a marked variation in synovial histologic appearances within the same joint. In clinical practice, synovial biopsy is most frequently performed to exclude or confirm a diagnosis of tuberculosis. However, chronic inflammatory granulation tissue with lymphocytes and epithelial cells may be all that is seen, with no tubercles, tuberculosis giant cells, or necrosis (50). This emphasizes the necessity for culturing biopsy specimens.

Electron microscopy and immunofluorescent studies are useful in research, but have no practical value as diagnostic procedures. Polymerase chain reaction and immunoelectron microscopy, which can detect small fragments of organisms, also have no value as diagnostic procedures, although they are used to confirm the diagnosis of Lyme arthritis (24,51).

Bone Biopsy

Bone biopsy is seldom required in rheumatology practice. Biopsy of bone tumors is best done by orthopaedic surgeons. Percutaneous transileal bone biopsy is also best done by experts in the field of bone and mineral metabolism. It is essentially a research tool, but is useful in diagnosis of osteomalacia when the results of radiographs and biochemical findings are ambiguous. The patient is premedicated with diazepam and meperidine (pethidine) and the biopsy is performed as an outpatient procedure. The site of biopsy is located 2 cm posterior to the anterior superior spine immediately inferior to the iliac crest. The physician or surgeon performing the procedure must be an expert in the technique because undue pressure to advance the trephine will crush osteopenic bone and create artifacts in bone with normal mineralization.

Complications are few, the most common being pain at the site of the procedure. Although hemorrhage is rare, it does tend to occur in patients with renal failure. The formation of a hematoma may cause temporary femoral nerve paralysis. Prior to the procedure, the patient is prescribed tetracycline 250 mg four times a day by mouth or demeclocycline 150 mg four times a day by mouth for three days on two occasions. After the first course, 14 days may elapse before the second course is given. The biopsy is performed five days after the second course is completed. The tetracycline or demeclocycline labels the bone, enabling histomorphometry of nondecalcified sections of trabecular bone. Tetracycline fluoresces with a light lemon color and demeclocycline produces an orange color.

The handling of the specimen from the biopsy requires special treatment, which most routine pathology laboratories are not equipped for because they do not have expertise in histomorphometry.

Muscle Biopsy and Other Investigations of Skeletal Muscle Disease

Muscle biopsy in rheumatologic practice is usually performed to confirm a diagnosis of polymyositis. The muscle chosen for biopsy should be clinically involved, but not severely wasted to the extent that it is extensively replaced by fatty and fibrous tissue. The biopsy should be taken from the belly of the muscle, not close to its tendinous insertion. The deltoid and quadriceps femoris muscles are probably the easiest and safest to biopsy because there is little danger of damaging major nerves or blood vessels.

Muscle biopsy can be performed using a Bergstrom muscle needle. This can be performed with a local anesthetic on outpatients. The major disadvantage is that the sample obtained is small. An open biopsy, however, provides a larger sample and can also be performed with a local anesthetic. A general anesthetic may be required in a child.

It is useful to perform an electromyography (EMG) before a muscle biopsy. This can provide guidance as to the appropriate site for biopsy. Because polymyositis and other muscle diseases a rheumatologist is likely to encounter are generally symmetrical, it is useful to perform EMG studies on only one side and use muscles on the opposite side for biopsy. A site at which an EMG needle has been inserted or a muscle has been recently injected should not be biopsied because artifactual changes can be confused with pathologic changes.

Diathermy should not be used when performing an open biopsy. Three cylinders of muscle, each approximataely 0.5–1.5 cm, should be removed rather than a single large specimen. The specimens should be quickly frozen for enzyme histology and biochemical analysis. The muscle cylinders can be fixed with two fine needles onto a piece of rubber before freezing. This ensures that contraction of the muscle fibers does not occur and enables the pathologist to prepare proper transverse, not slanted, sections. A small sliver of muscle can be obtained by the pathologist for electron miscroscopy.

Muscle, like all other tissues in the body, has only limited responses to disease. The changes that may be found in polymyositis, dermatomyositis, and inclusion body myositis have been well described by Drs. Anderson and Dalakas (52,53). Because polymyositis, dermatomyositis, and inclusion body myositis are focal disorders, a single biopsy may show no abnormality. Therefore, diagnosis has to depend on clinical findings, results of EMG studies, serum enzyme levels (especially creatine kinase), and anti-Jo-1 antibodies.

Although EMG changes are not entirely specific and do not necessarily correlate with severity of disease, they are helpful in excluding neurogenic disorders. The creatine kinase test is the most sensitive serum enzyme test currently available.

Other enzymes that also appear in muscle include aldolase, glutamic-oxaloacetic transaminase (GOT), aspartate aminotransferase (AST), glutamic-pyruvic transaminase (GPT), and alanine aminotransferase (ALT). Rarely are these enzyme levels elevated in the absence of an elevated creatine kinase.

Injury to muscle, such as an intramuscular injection or an EMG examination, results in elevation of serum creatine kinase. Strenuous exercise, surgery, and hypothyroidism also elevate serum creatine kinase. Perhaps the most common cause of elevation of serum creatine kinase is acute myocardial infarction. However, isoenzyme studies show that at least one-third of the total serum creatine kinase is made up of MB activity (e.g., the enzyme derived from myocardium in contrast to the MM activity derived from skeletal muscle). The situation is often less clear in clinical practice because at least 10% of patients with polymyositis and dermatomyositis have cardiac involvement, although it is usually asymptomatic. In addition, regenerating muscle contains the cardiac isoenzyme MB. Up to 20% of serum creatine kinase may be of this type, even in the absence of cardiac involvement (54). Serial determinations of serum creatine kinase are especially valuable in monitoring the response to corticosteroid or other therapy in polymyositis and dermatomyositis.

Myoglobinemia occurs in polymyositis and dermatomyositis. Myoglobinuria is also occasionally present. Although sequential determinations of serum myoglobin concentrations have been found useful in monitoring response to therapy, they are not often used. A bewildering array of nuclear and cytoplasmic antibodies have been described in patients with polymyositis and dermatomyositis. The Jo-1 antibody, which reacts with anti histidyl t-RNA synthetase, is present in 25% of patients with fibrosing alveolitis (55). However, the antibody is not found in patients with polymyositis associated with malignancy or dermatomyositis. Antibodies reacting with different components of skeletal muscle have been described, but are not used as routine tests. Patients with polymyositis and dermatocyositis frequently have positive rheumatoid and antinuclear factor tests and antibodies associated with overlap syndromes. Recently, Fraser et al reported that MRI may be useful in assessing disease activity and guiding sites for biopsy in inflammatory myopathies (56).

Myophosphorylase deficiency (McArdle's disease) causes progressive muscle weakness and can be easily misdiagnosed as polymyositis, especially as serum creatine kinase levels are elevated (57). A useful screening test is the forearm ischemic test. A venous blood sample is obtained for baseline lactate and ammonia concentrations. A sphygmomanometer is inflated to 20 mm Hg above systolic pressure. The patient squeezes a tennis ball every 2 seconds for 1.5 minutes. The cuff is deflated and venous blood samples are obtained at 1, 3, 5, and 10 minutes. Normally, lactate and ammonia concentrations increase by at least a factor of three over baseline values (58). Muscle biopsy in myophosphorylase deficiency shows deposits of glycogen, which can be demonstrated with PAS staining. Specific histochemical staining for myophosphorylase activity will confirm the diagnosis (59).

Sarcoidosis can involve a symmetrical proximal myopathy. Creatine kinase levels are frequently elevated. Noncaseating granulomas may be found on muscle biopsy, but lymphocytic infiltration, muscle necrosis, and regeneration may be all that is evident. Calcification of sarcoid granulomas may occur in patients who are hypercalcemic. MRI has proved useful in identifying muscle involvement with sarcoidosis; characteristic changes correlate with histologic changes (60).

Patients with hypothyroidism may complain of aches and pains, muscle cramps, and stiffness, which are exacerbated on exposure to cold. These may be the presenting symptoms of hypothyroidism. Proximal muscles are most commonly affected. Pinching or percussing with a tendon hammer results in a mounding phenomenon or myoedema, which is a pseudomyotonia consisting of a transient localized ridge or swelling. The muscles are usually of normal bulk, although hypertrophy has been described. In children, hypertrophy is known as Kocher-Debre-Semalaign syndrome and in adults it is known as Hoffman's syndrome. Muscle enzymes, including creatine kinase, are commonly elevated in hypothyroidism and the patient may be incorrectly diagnosed as having polymyositis. The creatine kinase may remain elevated up to 2 months after the patient becomes euthyroid. In primary hypothyroidism, the total thyroxine (T4) and free thyroid index (FT4I) are low, thyroid stimulating hormone (TSH) is high, and radioactive uptake (RAIU) is low. In Hashimoto's thyroiditis, the T4 and triiodothyronine (T3) are normal in the early stages of the disease, with a high TSH and slightly raised RAIU. In pituitary or hypothalamic hypothyroidism, the TSH is normal or low and the TSH stimulation test is normal. Certain drugs can induce myositis-like states. The rheumatologic agent that is most likely to do this is penicillamine (61).

Skin and Vessel Biopsies

Skin has the largest area of any organ and is the largest organ in the human body. Skin involvement in rheumatic disese is common. Arthritis may complicate a number of common skin diseases (e.g., severe acne vulgaris and psoriasis). Skin biopsies are useful in the diagnosis of progressive systemic sclerosis, systemic lupus erythematosus, and various forms of vasculitis. Biopsy can be performed by punch or scalpel. In most instances, only 4–5 mm of tissue obtained by punch biopsy is required. It is always best to select a fully-developed lesion and, if possible, to include a piece of normal tissue from the edge of the lesion. Although there are no contraindications to skin biopsy, areas where healing might be impaired should be avoided, as in ischemic skin. The technique of skin biopsy and the histologic findings in various dermatologic conditions have been well described (62–64).

In progressive systemic sclerosis, the histologic findings are identical to those in morphea. Normal histologic appearances may be present in skin that is clinically affected. Therefore, a negative biopsy does not exclude the diagnosis of progressive systemic sclerosis. Although the changes in chronic discoid lupus and systemic lupus erythematosus overlap, the latter has deposits of fibrinoid material. The lupus-band test involves direct immunofluorescence of both sun-exposed and involved and uninvolved skin. A positive test shows a characteristic fluorescent band at the dermal-epidermal junction, consisting of deposits of IgG or IgM and C3. Biopsies from sun-exposed, involved skin are negative in chronic discoid lupus, but positive in over three-quarters of patients with systemic lupus erythematosus. Approximately one-half of patients with systemic lupus erythematosus, especially those with lupus glomerulonephritis, have positive tests in biopsies taken from sun-protected areas of skin.

Amyloid may be diagnosed on skin biopsy or from aspirated subcutaneous fat. Although the diagnosis of psoriasis may be confirmed by skin biopsy, identical changes may be found in keratoderma blennorhagica in Reiter's disease. In rheumatoid arthritis, the histologic changes in an early subcutaneous nodule are similar to those found in rheumatic fever. However, central necrosis becomes prominent in rheumatoid nodules of 1 year or more duration. Subcutaneous nodules, similar to those found in rheumatoid arthritis, can occasionally be found in normal subjects. Gouty tophi usually do not require biopsy for diagnosis. Conditions such as febrile panniculitis (Weber-Christian disease) and eosinophilic fasciitis (Shulman's syndrome) require biopsy for diagnosis. Eosinophilic fasciitis requires a deep, wedge-shaped biopsy, which includes skin, subcutaneous tissue, fascia, and muscle.

Skin biopsy may help to differentiate the various vasculitides that affect skin. If sensory or motor symptoms are present in a patient suspected of having necrotizing arteritis, a sural nerve biopsy is helpful. Blind biopsies of clinically uninvolved tissues, such as skeletal muscle or testis, rarely yields positive confirmation of necrotizing arteritis. In the diagnosis of giant cell arteritis (temporal arteritis), it is necessary to take a segment of temporal artery 2–3 cm long because involvement may be patchy. If the biopsy is negative, some rheumatologists recommend that another be done on the opposite side.

Renal Biopsy

Renal biopsy is performed more often on patients with systemic lupus erythematosus than with any other rheumatic disease. Serologic markers of disease activity in systemic lupus erythematosus do not adequately reflect the degree of renal involvement. Tests of renal function and urinalysis do not discriminate irreversible lesions from active inflammatory changes. Therefore, renal biopsy findings are useful in selection of appropriate therapy and assessing prognosis.

An adequate renal biopsy should include at least eight glomeruli, in addition to renal tubules, interstitial tissue, and blood vessels. Similar to interpretation of radiographs of bones and joints, interpretation of renal biopsies should include a routine for assessing all four major components; glomeruli, tubules, interstitium, and blood vessels. The major abnormalities in systemic lupus erythematosus are found in the glomerulus and have been categorized into four degrees of severity: mesangial, mild proliferative, severe diffuse proliferative, and membranous (65). Immunofluorescent studies show Ig and C3 deposits along the glomerular basement membrane and in the mesangium and tubulointerstitial tissues. The extent of these deposits parallels the severity of the disease as defined by light microscopy. Electron microscopy can also be used to evaluate changes in lupus glomerulonephritis. Similar to immunofluorescent studies, electron microscopy provides little more useful information than that obtained by light microscopy.

Changes in the glomerulus that indicate acute or active changes include cellular proliferation (endothelial and mesangial cells) and polymorphonuclear leukocytes; fibrinoid necrosis and karyorrhexis (nuclear disruption and fragmentation) of degenerating leukocytes; cellular crescent formation in Bowman's space; and basement membrane thickening with broadened capillary loops (wire loops) and hyaline thrombi. In active lupus nephritis, edema and mononuclear cell infiltration are present in the interstitium. In chronic or inactive lupus nephritis, sclerosis of the glomerulus is present, which may be segmental, focal, or diffuse. The cellular crescents of active disease are replaced with connective tissue, becoming fibrous crescents. In the interstitium, there is fibrosis and the tubules become atrophied. A scoring system that has Kelvian merit has been developed for determining the activity of lupus nephritis (66).

Even in expert hands, needle biopsy of the kidney may result in complications. At least 5–10% of patients have complications, including gross hematuria. Massive hemorrhage may rarely occur and requires blood transfusion and arteriovenous fistulae. Continuing hemorrhage may require surgical intervention and even nephrectomy. Therefore, the clinician must decide whether biopsy of a kidney is justified to demonstrate vascular changes in progressive systemic sclerosis or secondary amyloidosis in rheumatoid arthritis. In progressive systemic sclerosis, changes are often present in patients who have no hypertension, proteinuria, or azotemia. In secondary amyloidosis, no treatment can be offered. Although the diagnosis of vasculitis can often be made on renal biopsy, it is important to try other less traumatic means (e.g., biopsy of a palpable nodular lesion in skin or sural nerve biopsy if peripheral sensory or motor complaints are present). Descriptions of the histologic findings in the various vasculitides can be found in several texts (67,68).

Biopsy of the Gastrointestinal Tract and Liver Biopsy

Salivary Gland

Minor salivary gland biopsy has proved useful in confirming a diagnosis of Sjögren's syndrome. Biopsy of the minor salivary glands on the inside of the lower lip or on the hard palate obviates the need for biopsy of the major salivary glands in Sjögren's syndrome (69). If done on the parotid gland, biopsy of a major salivary gland can result in facial nerve damage or lead to infection as a result of fistula formation. A minor gland biopsy should include at least eight glands. A focus is defined as consisting of at least 50 cells (70–72). In Sjögren's syndrome, local synthesis of IgM rheumatoid factor and IgG and IgM can be demonstrated by immunofluorescent methods, but is of little practical value. The finding of sialoadenitis cannot be considered diagnostic of Sjögren's syndrome because approximately one-quarter of patients with IgM rheumatoid factor-positive rheumatoid arthritis have positive biopsies, but no clinical evidence of the disease. There is no universally accepted definition of Sjögren's syndrome, although most experts require the presence of lymphocytic infiltration around labial salivary glands in biopsy specimens (72).

Positive biopsies may also be found in other connective tissue diseases. In progressive systemic sclerosis, the acinar glandular elements are reduced in size, presumably as a result of fibrosis. Parenchymal fibrosis increases with age. In our opinion, the real value of salivary gland biopsy is in making the diagnosis of other conditions that diffusely involve the salivary glands, such as sarcoidosis and lymphoma. Although major salivary gland biopsies are usually only done when a salivary gland tumor is suspected, some rheumatologists still favor biopsy of a parotid gland using a punch technique, which they claim is virtually free of side effects.

Gastrointestinal Biopsy

Biopsies of the gastrointestinal tract are usually performed to confirm a diagnosis of ulcerative colitis or Crohn's disease. Arthropathy has been described in patients with gluten-sensitive enteropathy, carcinoid tumors, and collagenous colitis, all of which may require biopsy for diagnosis. The diagnosis of Whipple's disease requires demonstration of PAS staining deposits in macrophages of the lamina propria of the small intestine. These deposits on electron microscopy reveal bacilliform bodies, which disappear on treatment with antibiotics. Similar changes can be found on synovial biopsy. Amyloid and necrotizing vascultis can frequently be diagnosed on rectal biopsy. In our opinion, the rectum is the site of choice for biopsy in both conditions.

Liver Biopsy

A liver biopsy should never be performed in a patient suspected of amyloidosis because massive hemorrhage can result. Liver biopsy is usually required to confirm a diagnosis of hepatitis or chronic biliary cirrhosis. In Felty's syndrome, nodular regenerative hyperplasia occurs, which may cause portal hypertension. This can occasionally occur in other connective tissue diseases and can easily be recognized by examination of biopsy specimens. In rheumatoid arthritis and other connective tissue diseases, mononuclear infiltrates of the portal tracts, hyperplasia of Kupffer cells, and sinusoidal dilation are common, but of no clinical significance. Salicylates commonly produce elevation of serum transaminase levels and, occasionally, serum bilirubin and alkaline phosphatase concentrations, which may be accompanied by mild hepatocellular degeneration or necrosis. However, these histologic changes are reversible. This transaminitis occurs with high concentrations of serum salicylate, especially in patients with hypoalbuminemia. In Reye's syndrome, which occurs even in the absence of aspirin therapy, characteristic but not pathognomonic microvesicular fatty infiltration may be seen on liver biopsy. Hepatic biopsy is helpful in Wilson's disease, but is seldom required to confirm the diagnosis of hemochromatosis.

Lung Biopsy

Lung biopsy is seldom required in rheumatologic practice because most rheumatic lung diseases can be diagnosed by other means. The most common reason for a lung biopsy is evaluation of a pulmonary nodule or coin lesion. Such nodules may occur in rheumatoid arthritis when other nodules, such as at the elbow, are present. A single coin lesion in a patient with seronegative non-nodular rheumatoid arthritis is almost certainly not a rheumatoid nodule, but more likely a tuberculoma or neoplasm. Biopsy of lung tissue is now more frequently performed in immunocompromised patients to diagnose the nature of infection.

Lung biopsy can be performed by percutaneous needle aspiration, transbronchial biopsy, and open lung biopsy. The methods of performing these procedures have been well described (73–75).

Percutaneous and transbronchial procedures are particularly useful in obtaining material for cytology and culture. Percutaneous biopsy can be used to biopsy peripherally situated lesions. The transbronchial approach is preferred for centrally placed lesions and infiltrative disease of the lungs. However, open biopsy is the most reliable method for obtaining adequate tissue. Lung biopsy involves risks, including hemorrhage and pneumothorax. Therefore, there must be good indications for performing a biopsy.

HEMATOLOGIC TESTS

Anemia is the rheumatologist's bête noire, especially the anemia of chronic disease in rheumatoid arthritis. The pathogenesis of this anemia is multifactorial (76,77). There is diminished absorption and transport of iron and a failure to release iron from iron stores. Erythropoietin is reduced and there is evidence of increased marrow resistance to its effects. Erythropoiesis and the production of erythroid progenitor cells are reduced. Although it seems likely that interleukin 1 (IL-1) plays a role in the pathogenesis of this type of anemia, the exact mechanism has not been identified. The anemia is normochromic, normocytic, and tends to parallel the severity of the inflammatory process, although not sufficiently to be useful as an outcome measure in clinical therapeutic trials of antirheumatic drugs. The hemoglobin concentration in the anemia of chronic disease seldom falls below 10 g/L in men and 9 g/L in women. The serum iron concentration is low and the total iron binding capacity (TIBC) is low or normal. Serum ferritin is often high due to the fact that this is an acute phase protein. Marrow iron stores are plentiful. The anemia does not respond to iron therapy but does respond to corticosteroids. Although recombinant human erythropoietin therapy causes improvement, the cost excludes its routine use.

In patients with rheumatoid arthritis, anemia is commonly due to iron deficiency. This is due to the fact that severely ill, crippled patients frequently eat poorly and have nutritional iron deficiency. In addition, patients may have chronic blood loss as a result of nonsteroidal antiinflammatory analgesic therapy. The source of bleeding is most often the stomach, but may also be from the lower small intestine (78). In iron deficiency anemia, the blood film shows a microcytic, hypochromic picture. The hemoglobin concentration is frequently below 9 g/L in females and 10 g/L in males. The serum iron concentration is low and the total iron binding capacity is normal or high. The serum ferritin concentration is normal or low and iron stores are depleted in the marrow. The two types of anemia often coexist, so that the distinction between them is blurred. In rheumatoid arthritis, iron deficiency anemia responds to oral iron therapy. A recent report by Vreugdenhill and colleagues suggests that the combination of the following three findings is virtually diagnostic of iron deficiency anemia in rheumatoid arthritis: mean corpuscular volume < 80 fl, serum ferritin < 50 ug/L, and serum transferrin > 50 μmol/L (77).

A macrocytic anemia is occasionally found in patients with rheumatoid arthritis and may be due to coexistant pernicious anemia, but is more likely due to folic acid deficiency. Folic acid deficiency is a consequence of inadequate dietary intake or a reaction to medication, such as sulfasalazine or methotrexate.

Hemolytic anemia occurs most commonly in systemic lupus erythematosus, with a warm IgG antibody Coombs positive test. Not all patients with a positive Coombs test become anemic and, occasionally, the Coombs test may be negative. Drug therapy can also cause hemolysis, especially in patients with glucose-6 phosphate dehydrogenase deficiency. The presence of Heinz bodies suggests the possibility of drug-induced hemolytic anemia. Aplastic anemia only occurs in patients with rheumatic disease as a result of drug therapy (e.g., azathioprine, cyclophosphamide, gold, methotrexate, penicillamine, and sulfasalazine).

Leukopenia (e.g., a total white cell count less than 1,500/mm^3) is common in a number of rheumatic diseases, including systemic lupus erythematosus, mixed connective tissue disease, and Felty's syndrome. Drugs such as phenylbutazone and injectable gold can also cause leukopenia and even agranulocytosis.

Leukocytosis occurs in a number of rheumatic illnesses, especially juvenile rheumatoid arthritis and the vasculitides. It should always alert the physician to the possibility of infection. Septic arthritis in adult rheumatoid arthritis usually is not associated with fever, rigors, or leukocytosis. In contrast, patients with systemic lupus erythematosus who are leukopenic and develop infection respond with a leukocytosis.

Eosinophilia occurs in a number of rheumatic diseases, including rheumatoid arthritis (especially when vasculitis is present), Sjögren's syndrome, eosinophilic myositis syndrome due to ingestion of L-tryptophan, sarcoidosis, and allergic granulomatous vasculitis (Churg-Strauss disease). Eosinophilia may occur in allergic reactions to drugs and may develop prior to the institution of drug therapy (e.g., gold therapy).

Thrombocytosis is a common feature of rheumatoid arthritis and the platelet count has been suggested as an outcome measure in clinical trials of antirheumatic drugs. The platelet count in patients with rheumatoid arthritis can occasionally be as high as $1,000,000/mm^3$. Thrombocytopenia is relatively common in systemic lupus erythematosus, occurring in approximately one-third of patients. It may antedate the onset of other features of the disease by a number of years. Therefore, patients with idiopathic thrombocytopenia should be screened for systemic lupus erythematosus. Thrombocytopenia also occurs in the mixed connective tissue disease syndrome and Felty's syndrome. It is frequently the result of antirheumatic drug therapy. When monitoring gold or penicillamine therapy, it is important to observe a fall in the platelet count, even when it is within the normal range, because it may be the first indication of severe thrombocytopenia or aplastic anemia.

Thrombocytopenia is extremely rare in Sjögren's syndrome. Hemorrhage as a result of thrombocytopenia usually occurs when the platelet count falls below $50,000/mm^3$.

Thrombotic thrombocytopenic purpura (Moschowitz syndrome) is a rare complication of systemic lupus erythematosus characterized by thrombocytopenia, microangiopathic hemolytic anemia (fragmented red cells on blood film), and widespread deposition of fibrin in small blood vessels as a consequence of activation of the clotting mechanism.

SERUM AND URINARY URIC ACID

There are many drugs that cause hyperuricemia (Table 2-1) (79). Secondary causes of gout most commonly encountered by the practicing doctor are those due to thiazide diuretics, renal failure, and myeloproliferative disorders (Table 5-4). Hyperuricemia is not synonymous with gout because many patients with elevated serum uric acid levels never develop gouty arthritis (80). Almost all uric acid in the body exists as monosodium urate. In urine, both uric acid and monosodium urate may exist (equal at pH 5.7). At low pH, uric acid predominates in urine. Therefore, the term urate is preferable when discussing serum. Arthritic symptoms should not be ascribed to gout just be-

Table 5-4. Causes of Hyperuricemia Other than Primary Gout

Purine overproduction

Myeloproliferative disorders
Polycythemia rubra vera
Hemolytic diseases
Psoriasis
Inherited enzyme defects (e.g., Lesch-Nyhan syndrome)
Glycogen storage disease (III, V, VII)

Decreased renal excretion

Renal failure
Lead nephropathy
Starvation
Diabetic ketoacidosis
Lactic acidosis
Hypothyroidism
Sarcoidosis
Toxemia of pregnancy
Bartter's syndrome

cause a high level of serum urate is found. A diagnosis of acute gout is best confirmed by the demonstration of urate crystals in synovial fluid. During an acute attack of gout, the serum urate concentration falls and may be within the normal range, but rises as the attack subsides. The likelihood of a patient developing acute gouty arthritis is dependent on the duration of hyperuricemia and the level of serum urate. The longer hyperuricemia lasts and the higher the level of serum urate, the more likely the patient will develop gout or renal stones. The serum urate concentration is elevated except when patients are receiving allopurinol or uricosuric therapy and during an acute attack of gout.

Serum and urinary uric acid are measured by an automated enzyme method based on the production of hydrogen peroxide when uric acid is oxidized by uricase (81). This method is more specific than the older colorimetric methods, which depend on the reduction of sodium tungstate and give slightly lower values by as much as 1 mg/dL (60 μmol/L). The upper limit of the normal range of serum urate is 7.0 mg/dL (420 μmol/L) in men and 6.0 mg dL (360 μmol/L) in women. The lower levels in adult women have been ascribed to the effects of estrogen on renal urate clearance. After menopause, serum urate concentrations rise to male values. Children have lower serum urate concentrations with no sex difference (4 mg/dL or 240

μmol/L). Serum urate concentration may be extremely low in early pregnancy. Synovial fluid concentrations of urate are the same as serum levels. Serum urate measurements should be converted from traditional (mg/dL) to SI units (μmol/L) by multiplying by 59.48. The multifaction factor for converting SI to traditional units is 0.0168.

Approximately two-thirds of the urate produced in the liver is eliminated in urine. The renal disposition of uric acid is complex, consisting of a four compartment system. Although glomerular filtration of uric acid is complete, 98–100% is reabsorbed in the proximal tubule. At least 50% of the reabsorbed uric acid is secreted back into the tubule, of which approximately 40–45% is again reabsorbed. Only 5–10% of the uric acid filtered by the glomerulus is excreted in the urine. These complex tubular reabsorption and secretion mechanisms explain the paradoxical effect of low and high doses of salicylates on urinary excretion of uric acid (Table 2-1). Low doses of salicylates (1–2 g/day) block uric acid secretion and high doses (3–5 g/day) block reabsorption. Probenecid inhibits tubular reabsorption of uric acid and prevents the decrease in uric acid secretion following low doses of salicylates. The paper by Rieselbach and Steeler gives a good account of the renal handling of uric acid by the kidney (82).

Uric acid in the urine is derived from endogenous production in the liver and exogenous dietary purines. One gram of dietary nucleic acid contributes approximately 150 mg of urinary uric acid. Thus, on an unrestricted diet, a normal subject may exceed 1000 mg/24 hr (600 μmol/d). After 5 days on a low purine (e.g., 10 mg/kg) or purine-free diet, the 24 hour urinary excretion of uric acid should not exceed 600 mg (360 μmol/d). Patients with primary or secondary gout who produce too much uric acid have higher amounts of uric acid in their urine.

It has been estimated that the risk of urolithiasis is one stone per year per 300 asymptomatic hyperuricemic persons and 1 in 100 per year in patients with gouty arthritis. Patients with secondary gout run twice the risk of developing kidney stones. The risk of uric acid stone formation is reduced with high fluid intake, alkalanization of the urine, and the avoidance of uricosuric drugs in patients who are overproducers of uric acid. The majority of patients with primary gout (approximately 90%) have a deficit in the renal handling of uric acid, so that overproduction and undersecretion contribute to hyperuricemia. Patients with excessive output of uric acid in the urine are prone to develop nephrolithiasis. Uric acid stones are usually small and round and, unless they contain calcium, radiolucent.

A high percentage of patients with primary gout also develop calcium oxalate and calcium phosphate stones, probably as a result of formation around a nidus of uric acid. In addition to nephrolithiasis, patients with primary gout also may develop what has been termed "gouty kidney" disease (83). This results from deposition of uric acid in the medullary interstitium, papillae, and pyramids. Chronic inflammatory changes may result, leading to nephrosclerosis, hypertension, and renal failure. There is controversy as to how often serious renal disease occurs in patients with primary gout. There is no controversy about the renal damage that can occur in patients with lymphoproliferative disorders who have high serum uric acid levels [e.g., 60 mg/dL (3568.8 mmol/L)] as a result of rapid nucleic acid breakdown with treatment. The complication can be prevented or reversed with allopurinol treatment.

SERUM PROTEINS

Hypoalbuminemia and elevation of the total serum globulins are characteristic findings in many inflammatory arthritides and connective tissue disorders. The level of serum albumin is seldom low enough to cause edema. Elevation of the total globulins may be associated with purpura hyperglobulinemia or hyperviscosity syndrome. Elevation of serum globulin is largely due to a rise in gamma globulin.

In recent years, it has been possible to distinguish five immunoglobulins comprising gammaglobulin: IgG, IgA, IgM, IgD, and IgE. In rheumatologic practice, measurement of these immunoglobulins is seldom necessary. IgA concentrations are elevated in ankylosing spondylitis, but do not predict clinical outcome or complications. Several patients with ankylosing spondylitis have developed an IgA nephropathy. Although measurement of IgE levels in patients with rheumatoid arthritis receiving gold therapy was used for awhile as a predictor of allergic reactions, subsequent studies failed to confirm the value of its estimation. Selective deficiency of serum IgA occurs in 1/500 persons to 1/2000 persons. Although 50% are free from symptoms, there is an increased frequency of rheumatoid arthritis, systemic lupus erythematosus, hemolytic anemia, thyroiditis, and type I diabetes in the remainder.

Congenital agammaglobulinemia can be diagnosed by the absence of immunoglobulins in the serum. The x-linked recessive form in boys may be complicated by arthritis. In adult practice, immunoglobulin assays are of particular importance in diagnosing monoclonal gammopathy due to an expanded clone of B cells. Thus, it is possible to diagnose multiple myeloma and Waldenström's macroglobulinemia. An increasing number of patients are being discovered with a monoclonal gammopathy who are free of symptoms over long periods of time (benign monoclonal gammopathy).

CRYOGLOBULINS

These are immunoglobulins that reversibly precipitate in the cold (4°C). They may be found in small amounts in normal subjects and in increased amounts in a variety of diseases. Cryoglobulins can be classified as type I (monoclonal), type II (mixed monoclonal and polyclonal), and type III (polyclonal).

Type I occurs in multiple myeloma and Waldenström's macroglobulinemia and is often associated with the hyperviscosity syndrome. Type II and type III occur in connective tissue diseases and infections. The polyclonal component is usually IgG and the monoclonal component is IgM with rheumatoid factor activity. Cutaneous vasculitis and glomerulonephritis are particularly associated with type II. Arteritis and neuropathies are associated with types II and III.

ACUTE PHASE PROTEINS

There are approximately 30 acute phase proteins synthesized by the liver as a consequence of inflammation or tissue damage. These proteins are produced in response to the liberation of cytokines, especially interleukins I (IL-1) and 6 (IL-6) and tumor necrosis factor (TNF). The cytokines are produced by monocytes and other cells, such as fibroblasts, at the site of inflammation or injury. They are responsible for the fever and leukocytosis that accompany inflammation and tissue destruction. At present, none of the cytokines are measured in routine practice. No correlation was found between serum levels of IL-6 and clinical indices of inflammation in chronic rheumatoid arthritis (84). The acute phase proteins that have been used as outcome measures in clinical therapeutic trials include C-reactive protein, haptoglobin, alpha-1–antitrypsin, fibrinogen, ceruloplasmin, alpha-1 acid glycoprotein (orosomucoid), complement components, plasminogen, and kinin polypeptides.

Elevation of C-reactive protein parallels increase in the erythrocyte sedimentation rate, with the exception of systemic lupus erythematosus, in which it is often low. However, should infection occur in a patient with systemic lupus erythematosus, the C-reactive protein will rise (85). Thus, measurement of C-reactive protein was considered useful in patients with systemic lupus erythematosus suspected of having an infection. In practice, however, a positive C-reactive protein has not proved to be a constant indicator of infection. Determination of C-reactive protein is of value in the differential diagnosis of benign monoclonal gammopathy, in which it remains negative, and malignant gammopathies. A further advantage of C-reactive protein is that it can rise 100 to 1000 fold, whereas other acute phase proteins tend to increase in the order of two to three fold at the most. C-reactive protein can be determined with great accuracy by nephelometry. Unlike the erythrocyte sedimentation rate, C-reactive protein levels do not increase with age (86).

The erythrocyte sedimentation rate, best done by the Westergren method, is not strictly an acute phase protein, although aggregation of erythrocytes is dependent on fibrinogen, alpha-2 macroglobulins, and immunoglobulins. Therefore, it can loosely be considered a composite acute phase reactant. As such, this simple and cheap test provides a useful general guide to the activity of the disease process and can alert the physician to the presence of possible pathology in a patient with vague symptomatology. Extremely high erythrocyte sedimentation rates are seen in malignancy, especially multiple myelomatosus, systemic lupus erythematosus, and polymyalgia rheumatica. However, very high levels (e.g., exceeding 100 mm/first hour) may occur in rheumatoid arthritis. The erythrocyte sedimentation rate is normal in noninflammatory joint disease. Mild inflammation, often present in osteoarthritis, may result in moderate elevation of 30–40 mm/first hour.

A normal erythrocyte sedimentation rate does not exclude disease. For example, we have seen several patients with polymyalgia rheumatica, in which the erythrocyte sedimentation rate is usually elevated, who had normal sedimentation rates. Rarely, the erythrocyte sedimentation rate will remain normal even in severe rheumatoid arthritis. The reason for this is not known.

The erythrocyte sedimentation rate is included in almost all trials of antirheumatic drugs. If it falls at all with nonsteroidal antiinflammatory analgesics, it is only by a moderate degree. The erythrocyte sedimentation rate often falls rapidly with corticosteroid therapy, especially if given in high doses. It is the only test currently available for monitoring response in polymyalgia rheumatica and temporal arteritis. With slow-acting antirheumatic agents, such as gold or penicillamine, the erythrocyte sedimentation rate shows a moderate decline after several months of treatment. Cyclosporine does not cause a reduction in the erythrocyte sedimentation rate in patients with rheumatoid arthritis (87). It has been suggested that drugs that control the erythrocyte sedimentation rate and C-reactive protein in rheumatoid arthritis also reduce radiologic progression in rheumatoid arthritis (88).

Recently, plasma viscosity has been suggested as an alternative to the erythrocyte sedimentation rate, especially as an outcome measure in clinical therapeutic trials (89).

SERUM CALCIUM AND PHOSPHORUS

Perhaps the most useful biochemical estimation in clinical medicine is the determination of serum calcium. There are many causes of hypercalcemia (Table 5-5) and hypocalcemia (Table 5-6). Serum calcium consists of three parts; protein (90% albumin) bound (40%); ionized (48%); and complexed (to phosphate, citrate, and bicarbonate) [12%]. Ionized calcium is the component important for physiologic processes (e.g., it is the active form), but

Table 5–5. Causes of Hypercalcemia

Primary hyperparathyroidism
Malignancy
    Lytic bone metastases
    Humoral peptide of malignancy
    Ectopic production of 1, 25-dihydroxyvitamin D
    Other factor(s) produced ectopically
Nonparathyroid endocrine disorders
    Thyrotoxicosis
    Pheochromocytoma
    Adrenal insufficiency
    VIPoma
Vitamin D
    Vitamin D toxicity
    Granulomatous diseases
        Sarcoidosis
        Tuberculosis
        Histoplasmosis
        Coccidioidomycosis
        Leprosy
    Lymphoma
Medications
    Thiazide diuretics
    Lithium
    Estrogens/antiestrogens
    Milk-alkali syndrome
    Vitamin A toxicity
    Vitamin D toxicity
Familial hypocalciuric hypercalcemia
Immobilization
Parenteral nutrition
Acute and chronic renal disease

Reproduced with permission from: Bilezikian JP. Hypercalcemic states, their differential diagnosis and acute management. In: Coe FD, Favus MJ, eds. Disorders of bone and mineral metabolism. New York: Raven Press, 1992; 492.

Table 5–6. Causes of Hypocalcemia

Chronic Hypocalcemia
  Hypoparathyroidism
    Surgically induced
    Idiopathic
      Early
        DiGeorge syndrome (with thymic aplasia)
        MEDAC/HAM*
        Isolated persistent neonatal
      Isolated late-onset
    Functional
      Hypomagnesemia
      Neonatal (e.g., maternal hyperparathyroidism)
    Infiltrative
      Hemosiderosis
      Wilson disease
      Secondary neoplasia
  Pseudohypoparathyroidism
  Osteomalacia
    Decreased bioavailability of vitamin D
      Decreased UV light
      Nutritional
      Nephrosis
      Malabsorption
    Abnormal metabolism of vitamin D
      Chronic renal failure
      Vitamin D-dependent rickets type I (aut. rec.)
      Anticonvulsants
    Abnormal response to vitamin D
      Vitamin D-dependent rickets type II
      Malabsorption (celiac/Crohn's/short bowel)
Acute/Subacute Hypocalcemia
    Severe illness
      Septicemia (gram-negative)
      Fat embolism
      Pancreatitis
      Tumor lysis
      Rhabdomyolysis
      Excess phosphate administration
      Toxic shock syndrome
    After neck surgery
    Osteoblastic metastasis
    Drugs (e.g., calcitonin, mithramycin, citrated blood)

*MEDAC-multiple endocrine deficiency-autoimmune-candidiasis; HAM-juvenile familial endocrinopathy-hypoparathyroidism-Addison disease-moniliasis.
(Adapted with permission from: Eastell R, Heath H III. The hypocalcemic states, their differential diagnosis and acute management. In: Coe FD, Favus MJ, eds. Disorders of bone and mineral metabolism. New York: Raven Press, 1992; 572.)

for clinical purposes it is the total serum calcium that is usually measured. Because 40% of the total serum calcium is protein bound, low values should be expected in patients with hypoalbuminemia. There is a normal diurnal variation in serum calcium, with peak to nadir differences of 0.5 mg/dL (0.13 mmol/L). Prolonged upright posture and venostasis causing hemoconcentration can also increase serum calcium concentrations. The normal values in adult males are 9.04–10.30 mg/dL. In adult females, the normal values are 8.93–10.50 mg/dL. To convert these values to mmol/L, divide by 4. The serum calcium concentration can be adjusted to the serum albumin concentration according to the following equation:

$$\text{Total calcium concentration (mg/dL)} + 0.8 \times [4\text{-albumin concentration (mg/dL)}]$$

Phosphorus exists in plasma in an organic form (e.g., phospholipids) and an inorganic form. In clinical practice, only the inorganic form is measured, of which 15% is protein bound. Although the term phosphate is often used for phosphorus, it is incorrect. Its measurement is useful in investigation of metabolic bone disease.

RENAL FUNCTION

Assessment of renal function is important in diseases in which the kidney is involved, such as systemic lupus erythematosus. It is also important in diseases such as osteoarthritis in the elderly because nonsteroidal antiinflammatory analgesics may precipitate acute renal failure. Although the serum creatinine concentration is most often used to assess glomerular filtration rate, it is not ideal because creatinine is also actively secreted in the tubules. The assay of serum creatinine is subject to interference from noncreatinine chromogens. Serum creatinine may rise after a meal of cooked meat or exercise. Both serum creatinine and creatinine clearance are reduced in the elderly as a result of lean body mass. The correcting formula of Cockroft and Gault for body weight is (92):

$$\text{Creatinine Clearance} = \frac{(140\text{-age}) \times \text{body weight (kg)}}{0.81 \times \text{serum creatinine } (\mu\text{mol/L})}$$

In women, the value should be reduced by 15%. Even with this correction, measurement of creatinine clearance is subject to considerable error (e.g., as much as 27% in a recent study) (93). Therefore, most clinicians only perform serum creatinine determinations to assess glomerular filtration rate.

Due to reduction in lean body mass in the elderly, the serum creatinine concentration, which would otherwise be elevated, may be in the normal range. In addition to exercise and the consumption of meat, elevation of serum creatinine may also occur with drugs, such as salicylates and cimetidine, which block tubular secretion. Plasma urea is less valuable as an index of glomerular filtration rate than serum creatinine because it is influenced by a number of factors, especially dehydration.

Urinalysis

Microscopic hematuria (e.g., more than 3 red blood cells per high power field) may occur in a large number of diseases involving the urinary tract. In rheumatology practice, it is most frequently found in connective tissue disease, especially systemic lupus erythematosus, and the various forms of vasculitis. Microscopic hematuria may also be found after initiation of salicylate therapy, especially acetyl salicylic acid, although this usually clears in 10 to 14 days. Microscopic hematuria may herald the onset of gold or penicillamine nephrotoxicity. Gross hematuria may be seen in cyclophosphamide-induced cystitis, uric acid lithiasis, and polyarteritis nodosa.

Red cell casts are indicative of glomerulitis. They are fragile and may not be seen when there has been a delay in examining the urine specimen. White cell casts may occur in the urine of patients with systemic lupus erythematosus, but should always prompt investigation for infection. In the nephrotic syndrome, waxy-appearing casts are common. In renal amyloidosis, these can be shown to consist of amyloid material.

Proteinuria is usually assessed on 24 hour specimens, and is categorized as normal (< 0.15 q/24 hours), mild (> 0.15 and < 0.5 q/24 hours), moderate ($0.5–3.5$ g/24 hours), and severe (> 3.5 g/24 hours). The severe category is in the nephrotic range.

Proteinuria is a feature of the connective tissue diseases, such as systemic lupus erythematosus, progressive systemic sclerosis, the various vasculitides, and amyloidosis. Gold and penicillamine therapy may cause glomerular damage, leading to proteinuria. Nonsteroidal antiinflammatory agents may also lead to proteinuria as a result of interstitial nephritis. The metabolites of tolmetin sodium (Tolectin) give false positive tests for proteinuria using any test that relies on acid precipitation as its end point (e.g., sulfosalicylic acid). No interference occurs with Albustix Reagent Strips.

Although proteinuria has no diagnostic specificity, its appearance in a patient with progressive systemic sclerosis may herald the onset of an acute hypertensive crisis. Bence-Jones proteinuria consists of light chains of IgG and is the distinctive feature of multiple myeloma. Light chains of IgG may also be found in Waldenström's macroglobulinemia. Bence-Jones proteins coagulate on warming and disappear on boiling of the urine. Today, they are identified by precipitation in 50-fold concentrated urine using antilight chain antibodies. Patients with multiple myeloma may develop heavy proteinuria, which is usually due to beta-2 microglobulin amyloid deposition in the glomerular basement membrane.

LIVER FUNCTION TESTS

Liver disease is frequently complicated by arthritic manifestations. It occurs in a number of arthritic and rheumatic diseases and as a consequence of therapy. For example, polyarthritis may occur in chronic biliary cirrhosis and bony erosions may develop as a consequence of xanthomatous deposits. Chronic active hepatitis may be associated with lupus-like manifestations and positive LE cells (lupoid hepatitis). Viral B hepatitis is commonly associated with inflammatory polyarthritis, urticaria, and rashes, especially during acute phases of the illness. Necrotizing arteritis may complicate viral B hepatitis. Polyarthritis and erythema nodosum have been reported after administration of recombinant hepatitis B vaccine. Hepatitis A infection has only rarely been reported to be associated with arthritis and vasculitis. Liver transplantation may cure hypertrophic osteoarthropathy associated with liver disease or it may be followed by the development of hypertrophic osteoarthropathy shortly after the operation. Hemochromatosis and Wilson's disease may be complicated by joint disease and chondrocalcinosis.

Patients with rheumatoid arthritis frequently have elevated serum alkaline phosphatase and transaminase levels. These may be due to drug therapy or histologic changes in the liver consisting of mononuclear cell infiltration of the sinusoids, portal tracts, and Kupffer cell hyperplasia. Similar findings are common in Still's disease. In a small proportion of patients, antimitochondral antibodies may be found, which are associated with chronic biliary cirrhosis. This liver disease occurs with increased frequency in patients with Sjögren's syndrome. In Felty's syndrome, liver function tests are frequently abnormal and associated with nodular regenerative hyperplasia. The serum alkaline phosphatase may be elevated as a result of bone disease and liver abnormalities. Measurement of bone and liver isoenzymes of alkaline phosphatase is necessary to determine the source. In liver disease, high levels of serum alkaline phosphatase are especially found in chronic biliary cirrhosis. In bone disease, Paget's disease causes extremely high alkaline phosphatase concentrations. Measurement of the enzyme in this disease is particularly valuable in monitoring activity and the effects of treatment. In giant cell arteritis, the serum alkaline phosphatase is frequenty elevated, especially in patients who have constitutional symptoms. It is seldom elevated in polymyalgia rheumatica.

Antirheumatic drug therpay can cause elevation of the serum transaminases, especially high doses of salicylates; hence the term aspirin transaminitis. The elevation is disease nonspecific and appears to be common in patients with reduced serum albumin levels, which presumably lead to higher concentrations of free salicylate. Occasionally, there may be elevation of the serum bilirubin and serum alkaline phosphatase in such patients. All these biochemical abnormalities quickly reverse with discontinuation of therapy. Progression to serious liver disease has not been reported.

The hepatocerebral syndrome described by Reye was associated with the use of aspirin in children during epidemics of influenza and varicella. It did not appear to be due to salicylate intoxication and the incidence of the complication has fallen since acetaminophen (paracetamol) has replaced aspirin for the treatment of fever in children. Reye's syndrome has been reported in children with juvenile rheumatoid arthritis and other connective tissue diseases. Biochemically, the illness is characterized by elevation of liver enzymes and ammonia and a prolonged prothrombin time.

Nonsteroidal antiinflammatory analgesics may occasionally produce hepatocellular necrosis and steatosis and, rarely, cholestatic jaundice. It is recommended by some rheumatologists that regular liver function tests be performed on patients on long-term therapy with these drugs, especially elderly patients. We are not convinced of the necessity of this or whether it would make it possible to prevent such a complication. Phenylbutazone appears to be more likely to cause this complication than the other nonsteroidal antiinflammatory agents. A review of these and other rare side effects to nonsteroidal antiinflammatory agents has been publshed by O'Brien and Bagby (94).

Hepatotoxicity occurring with gold and penicillamine therapy is rare. It is not certain whether it is due to intercurrent viral hepatitis or the effects of the drugs. Azathioprine therapy may be associated with elevation of serum transaminases, but these quickly return to normal when treatment is stopped. Methotrexate therapy in patients with rheumatoid arthritis is well recognized as producing abnormal liver function tests. Chronic administration can lead to fibrosis. However, liver function tests do not predict significant liver damage. The risk factors appear to be obesity, alcohol consumption, and the total dose and duration of therapy. Hepatotoxicity can occur in patients treated with sulfasalazine and cyclosporine. Allopurinal can also cause severe hepatic failure, especially in patients with chronic renal failure.

It is clear that routine liver function tests have many uses in rheumatologic practice.

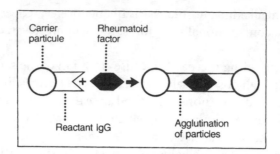

Figure 5-2.   Principle of rheumatoid factor tests.

RHEUMATOID FACTOR TESTS

Rheumatoid factors are autoantibodies to antigenic determinants of the Fc portion of IgG molecules. The most commonly used tests for IgM rheumatoid factors are the bentonite and latex agglutination tests and the Waaler-Rose or sheep-cell agglutination test (Fig. 5-2). In the bentonite and latex agglutination tests, human IgG acts as antigen. Rabbit IgG is used in the sheep-cell agglutination test. Other tests that use human IgG and have proved sensitive include the radioimmunoassay (RIA) and enzyme-linked immunosorbent assay (ELISA). Recently, automated nephelometric methods have become popular. These methods use aggregated IgG as antigen. The light scattered from particles suspended in solution as a result of formation of IgG-IgM rheumatoid factor complexes forms the basis of measurement. The results of RIA, ELISA, and nephelometry tests are expressed in International Units per millimeter (IU/ml). Because IgM is an excellent agglutinator, latex nephelometry primarily detects IgM rheumatoid factor. Turbidity-based nephelometry has the potential to detect IgM, IgG, and IgA rheumatoid factor activity (95).

Radioimmunoassay, ELISA, and turbidity-based and latex-based nephelometry procedures have a high degree of precision and sensitivity. In the next decade, these procedures can be expected to replace the current agglutination methods. There is a high correlation in results between these methods. The relation between latex agglutination titers and nephelometry results found in one of our laboratories is summarized in Table 5-7.

In the rheumatic diseases, rheumatoid factors exhibit considerable immunochemical heterogeneity. In addition to IgM rheumatoid factor, which is detected by the standard agglutination tests, IgG, IgA, and IgE rheumatoid factors have been identified. In rheumatoid arthritis, rheumatoid factor cross-reacts with several mammalian IgGs, including rabbit, bovine, and

Table 5–7. Relation Between Latex Agglutination Titers and Nephelometry Units in Tests for Rheumatoid Factor

Latex agglutination titers	Nephelometry (IU/ml)
Negative	Out of range, low
160	60–70
320	70–140
640	140–280
1280	280–560
2560	560–1120
5120	1120–2240
10240	2240–4480
20480	4480–8960

Courtesy of Mrs. B. Roberge, Chief Technologist, Department of Immunology and Histocompatability, McMaster University Faculty of Health Sciences, Hamilton, Ontario, Canada.

equine. Although rheumatoid factor reacts especially with human IgG, exceptions have been reported. In rheumatoid arthritis, rheumatoid factor is produced by B cells in the synovium. Polyclonal B cell activators, including Epstein-Barr virus, are potent inducers of rheumatoid factor. Altered glycosylation of IgG in rheumatoid arthritis has been demonstrated and the resulting altered structure of the molecule has been considered a possible cause of production of rheumatoid factor. Similar perturbations of glycosylation of IgG have been observed with aging and may explain why healthy elderly persons have positive tests for rheumatoid arthritis.

Rheumatoid factor is positive in many chronic inflammatory and infectious diseases. High titers are found particularly in rheumatoid arthritis and Sjögren's syndrome (Table 5-8). Persistently high titers in rheumatoid arthritis usually suggest a poor prognosis and are frequently associated with vascular and granulomatous complications (96). However, rheumatoid factors transfused into normal subjects are without ill-effect (97). Rheumatoid factors just released from B cells are highly avid and specific for human IgG and form complexes, which are quickly cleared from the circulation (98). Patients with rheumatoid arthritis with high titers of circulating rheumatoid factors do not contain these avid rheumatoid factors. This presumably explains why their transfused plasma is without ill effect.

Tests emphasizing human IgG are more sensitive than those using rabbit IgG, with the exception of Kala Azar. Tests for rheumatoid factor may remain negative during the first year of rheumatoid arthritis. A proportion of patients remain seronegative for the entire course of their illness. These seronegative

Table 5–8. Diseases Associated with Positive Tests for IgM Rheumatoid Factor

Arthritis and connective tissue diseases
   Rheumatoid arthritis*
   Juvenile rheumatoid arthritis[†]
   Systemic lupus erythematosus
   Progressive systemic sclerosis
   Dermatomyositis and polymyositis
   Sjögren's syndrome*
   Seronegative spondyloarthropathies[†]
Chronic bacterial and spirochetal infections
   Subacute bacterial endocarditis
   Syphilis
   Tuberculosis
   Leprosy
   Lyme disease
Viral diseases
   Infectious mononucleosis
   Rubella
   Cytomegalovirus infection
   Influenza
   Acquired immune deficiency syndrome
Chronic inflammatory diseases of unknown etiology
   Chronic
   Chronic interstitial pulmonary fibrosis
   Sarcoidosis
Miscellaneous
   Waldenström's macroglobulinemia
   Polyclonal cryoglobulinemia
   Parasitic infections
   Asbestosis
   Silicosis
   Periodontal disease
   Immunization with bacterial and viral antigens
   Elderly subjects

*Often in very high titers.
[†]Rheumatoid factor tests negative in patients with onset <5 years, 5–10% positive in disease onset 5–10 years, and gradually reaching adults levels in those developing disease between 10 and 15 years.
[†]Despite the adjective, between 10–15% of patients have positive tests, but in low titer.

patients tend to have milder disease and rarely develop systemic complications. Patients who are seronegative for rheumatoid factor according to classical tests may have IgG and IgM antiglobulins (99,100). Aho et al recently showed that rheumatoid factors detected by classical methods may antedate rheumatoid arthritis (101). Patients with primary Sjögren's syndrome frequently have high titers of rheumatoid factor detected by human and rabbit IgG methods, but do not develop rheumatoid arthritis for unknown reasons.

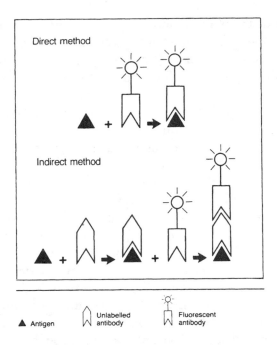

Figure 5-3. Principle of antinuclear factor test.

ANTINUCLEAR AND ANTICYTOPLASMIC ANTIBODIES

Autoantibodies in systemic lupus erythematosus and other connective tissue diseases react with constituents of cell nuclei, such as nucleic acids (DNA, RNA) and histones, and cytoplasmic constituents, such as mitochondria and ribosomes. The autoantibodies are commonly detected by indirect immunofluorescence (Fig. 5-3) and immunodiffusion when they are directed to soluble cellular macromolecules (e.g., anti-Ro, anti-La, and anti-Sm). Four different patterns of antinuclear antibodies are demonstrated by indirect immunofluorescence. A homogeneous pattern is suggestive of anti-DNA-histone antibodies. The rim or shaggy pattern usually indicates the presence of anti-DNA antibodies: a speckled pattern with a variety of antibodies, including anti-Sm, anti-ribonucleoprotean (UIRNP), anti-Ro, anti-La, and anti-centromere antibodies. The latter antibodies produce a characteristic discrete speckled pattern. Nucleolar staining denotes anti-RNA antibodies. However, recent data suggests that homogeneous and speckled patterns may be caused by any number and combinations of specific antibodies (102–104). It now seems that with the exception of anti-centromere antibodies, the pattern of immunofluorescent nuclear staining is a poor predictor of antibody specificity.

The result of the immunofluorescence antinuclear antibody test is of much greater importance. Although the test has a low diagnostic specificity for systemic lupus erythematosus (less than 50%), a negative test is useful in excluding the diagnosis (the likelihood being less than 5%). The major antigenic determinants reside on the pentose-phosphate backbone of the double-stranded DNA molecule (103). Single-stranded DNA manifests many different epitopes, which is the reason why antibodies are commonly found in conditions other than systemic lupus erythematosus. Therefore, tests for antibodies to single-stranded DNA are of no clinical value.

Anti-DNA antibodies are detected by the Farr technique or the Crithidia immunofluorescent test. The specificity of the Crithidia immunofluorescent test is higher than the Farr technique, but the sensitivity is lower. Antibodies to double-stranded DNA are found in patients with systemic lupus erythematosus and, rarely, in other diseases. There is reasonably good correlation between titers of anti-DNA antibodies and activity of systemic lupus erythematosus, although exceptions do occur. Anti-Sm antibodies are highly specific for systemic lupus erythematosus when tested by immunodiffusion, although they are only found in 20% of patients.

Although anti-n(nuclear)RNP antibodies occur in many connective tissue diseases, high titers are found almost exclusively in patients with mixed connective tissue disease. Anti-Sm and anti-nRNP antibodies share partial identity in Ouchterloney immunodiffusion. Virtually all anti-Sm antibody-containing sera have anti-nRNP antibodies.

Anti-Ro (SS-A) and anti-La (SS-B) antibodies are determined by immunodiffusion, counterimmunoelectrophoresis, or enzyme-linked immunoassay. They are found in a high percentage of patients with primary Sjögren's syndrome and in patients with systemic lupus erythematosus and neonatal lupus syndrome. Infants born of mothers with these antibodies, especially anti-Ro, have an increased incidence of congenital heart block. There appears to be a genetic link between these antibodies and HLA-DR3.

Anticentromere antibodies, detected by indirect immunofluorescence, are most commonly associated with the CREST syndrome (calcinosis, Raynaud's phenomenon, esophageal dysmotility, sclerodactyly, and telangiectasia) and primary Raynaud's syndrome. Anti Scl-70 (Topoisomerase 1) antibodies are found especially in progressive systemic sclerosis.

In primary biliary cirrhosis, mitochondrial antibodies are present in high titer. They react with an antigen in mitochondria and, as a consequence, interfere with assessment by indirect immunofluorescence of other anticytoplasmic antibodies. Smooth muscle antibodies are directed against actin and are found especially in patients with chronic active hepatitis.

Antibodies that react with proteins in the primary granules of neu-

Table 5–9. Autoantibodies in Connective Tissue Diseases

Antigen	Clinical association of autoantibody
Native or double-stranded DNA	50–60% SLE (not drug-induced) 5% chronic active hepatitis Rarely other connective tissue diseases

Histones	70% SLE
	H2A/H2B Ig and IgM autoantibodies in pro-
	  cainamide-induced SLE
	H2A/H3 IgM autoantibodies in hydralazine-
	  induced SLE

Sm	30–40% SLE
UIRNP	30–40% SLE 95–100% MCTD
Ro (SSA)	30–40% SLE 60–70% SS 10% normal, but very low titer
La (SSB)	15% SLE 25% SS Both anti-Ro and anti-La antibodies associated with neonatal SLE and congenital heart block
Topoisomerase 1 (Scl-70)	20–30% PSS
Centromere (Kinetochore)	20–30% PSS
Histidyl-tRNA Synthetase (Jo-1)	25–30% PM, especially with interstitial lung disease
Neutrophil lysozomal enzymes (ANCA)	60–70% Wegener's granulomatosis

SLE = systemic lupus erythematosus
MCTD = mixed connective tissue disease
SS = Sjögren's syndrome
PSS = progressive systemic sclerosis
PM = polymyositis

trophils and lysosomes of monocytes have recently been described. These anti-neutrophil cytoplasmic antibodies (ANCA) are found in patients with vasculitis, especially Wegener's granulomatosis.

The relationships of these antibodies to nonhistone nuclear proteins, RNA-protein complexes, and cytoplasmic antigens to various diseases are summarized in Table 5-9. The LE cell test detects antibody to DNA-histone and is now obsolete.

Lymphocytotoxic antibodies are directed against antigenic determinants on cell surfaces. They participate in complement-dependent lysis of lymphocytes; hence, their name. These antibodies are present in systemic lupus erythematosus, juvenile rheumatoid arthritis, and Sjögren's syndrome. They are associated with lymphopenia and, in systemic lupus erythematosus, with central nervous system complications. Tests for lymphocytotoxic antibodies are complex and time-consuming and not part of routine investigation.

ANTIPHOSPHOLIPID ANTIBODIES

Anti-DNA antibodies cross react with cardiolipin. This results in the false positive complement-fixation Wassermann and flocculation venereal disease research laboratory (VDRL) reactions (105). Lupus anticoagulant and anticardiolipin antibodies, collectively known as antiphospholipid antibodies, are found in patients with systemic lupus erythematosus; other autoimmune, infectious, and neoplastic diseases; and in patients receiving certain drugs (e.g., hydralazine) (106). These antibodies are immunoglobulins of the IgG or IgM class, which bind to phospholipids (e.g., cardiolipin phosphatidyl serine) and interfere with prothrombin activation. As a result, they are associated with arterial and venous thrombosis, repeated abortion, and thrombocytopenia (the triad comprising the antiphospholipid syndrome) (107). Even in the absence of systemic lupus erythematosus or any other disease, the lupus anticoagulant is associated with thrombosis.

DRUG-INDUCED LUPUS

Approximately 60 drugs have been implicated in causing a lupus-like syndrome. However, only five (hydralazine, procainamide, isoniazid, methyldopa, and chlorpromazine) have been recognized as definitely causing the syndrome, with another three (phenytoin, penicillamine, and quinidine) considered possible causes. Antibodies to histone are present in the syndrome, especially antibodies against H2a and H2b histone fractions. Neurologic and renal complications do not occur and the clinical and laboratory abnormalities quickly disappear when the medication is stopped. Patients who are slow acetylators are particularly prone to develop this complication, especially with hydralazine and procainamide (108). In active systemic lupus erythematosus, 80% of patients have antibodies to histone. Only 50% who are inactive or only mildly active have antibodies. However, antibody titers are higher in drug-induced lupus and are directed against H2a and H3b histone fractions.

Table 5–10. Inherited Complement Deficiencies

Component	Associated diseases
$C1_q$, $C1_r$, $C1_s$, C4	SLE, GN, pyogenic infections†
C2*	SLE, GN, infections
C3	SLE, GN, severe immune deficiencies
C5	SLE, meningococcal and monococcal infections
C6, C7, C8	Meningococcal infection; rarely SLE
C9†	Meningococcal infection
Factor I	Pyogenic infections
Factor H	Hemolytic uremic syndrome
Properdin	Pneumonia, meningococcal infection

*50% healthy
†Common in Japanese, who apparently remain healthy
†SLE-systemic lupus erythematosus; GN-glomerulonephritis
(Adapted with permission from: Morgan PB. Complement clinical aspects and relevance to disease. London: Academic Press, 1990.)

Immune Complexes

High concentrations of circulating immune complexes are often detected in patients with vasculitis, systemic rheumatic syndromes, infections, and some neoplasms. However, there is no clear relationship of such complexes with pathologic changes. Consequently, assays are of little practical value in diagnosis or monitoring therapy.

Complement

The complement cascades, the classical and alternative pathways, are an extremely complex system. The importance of the cascades is clearly due to the fact that inherited deficiences of particular protein components predispose to certain diseases, as summarized in Table 5-10. The classical pathway is activated by immune complexes and the alternative pathway is promoted by components of microorganisms (Fig. 5-4). Both result in splitting C3, which is the central event leading to opsonization of bacteria and viruses, target cell killing, and macrophage activation (Fig. 5-4). The system is so complicated that it has been referred to by Morgan in his excellent book as the most "unloved" (109).

In rheumatology practice, measurement of total complement (CH50) and C3 and C4 levels is confined to systemic lupus erythematosus and immune-mediated illnesses. Complement activation occurs when immune

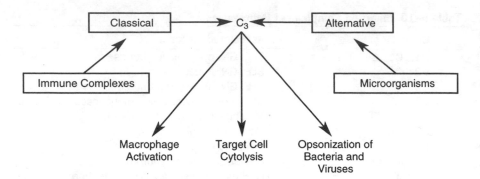

Figure 5-4. Activation of the complement system.

complexes are deposited in tissues, such as occurs in lupus glomerulonephritis with anti-DNA-DNA complexes. Thus, monitoring of CH50 and C3 and C4 levels is important and, with DNA antibody titers, provides useful monitoring of disease activity in patients with systemic lupus erythematosus. There are several excellent reviews of antinuclear and anticytoplasmic antibodies, complement, and immune complexes (110–112).

HLA Antigens

The human leukocyte antigen (HLA) system is located on the short arm of chromosome 6. The genes are grouped into three classes. Class I includes HLA-A, B, and C expressed on all nucleated cells and platelets. Class II includes D region molecules that are grouped into three major subregions known as DR, DQ, and DP and are expressed on cells of the immune system, including B cells, monocytes, dendritic cells, and stimulated T cells. The class III molecules are not part of the HLA system. They comprise the genes determining the third and fourth components of the classical complement pathway and alternative complement pathway, and properdin factor B. Other genes in the class III group determine 21-hydroxylase, human tumor necrosis factor, lymphocytoxin, and heat shock proteins.

Class I Antigen-Associated Disease (113)

In Europe and the United States, approximately 90% of Caucasian patients with ankylosing spondylitis are HLA-B27 positive compared to 6–8% of the normal population. In contrast, black Americans, whose frequency of B27 is only 2–3%, develop ankylosing spondylitis less often and with a weaker association of approximately 50% with B27. There is a relationship between the frequency of B27 in different races and the incidence of ankylosing spondylitis. The incidence of ankylosing spondylitis is low in Africans and Japanese and high in certain American Indians, especially the Haida of Queen Charlotte Islands in British Columbia. The risk of developing ankylosing spondylitis in B27 positive individuals has been calculated as only 2%, but increases to 25% in B27 positive relatives of patients with the disease. Patients who develop uveitis and cardiac complications of ankylosing spondylitis are B27 positive. These complications are rarely seen in B27 negative patients. Patients who are B27 positive, especially if homozygous, are more likely to have severe ankylosing spondylitis. The diagnosis of ankylosing spondylitis is based on clinical and radiologic criteria. A positive B27 does not establish or confirm the diagnosis, but merely provides supportive evidence. The HLA-B27 test is of no value in routine screening of patients with back pain.

Patients with Reiter's syndrome and other forms of reactive arthritis have a high incidence of B27 positive tests (approximately 75%). Patients with acquired immunodeficiency syndrome who are B27 positive are particularly liable to develop severe Reiter's syndrome. HLA-B27 is associated with ankylosing spondylitis in psoriatic arthritis (50% of patients) and inflammatory bowel disease (30–50% of patients).

Class II and Class III Antigen-Associated Diseases (114)

A strong association has been found between DR4 and adult seropositive rheumatoid arthritis in Caucasians, blacks, and Japanese. This association has not been found in Arabs, Jews, and Yakiima Indians. There is evidence that HLA-DR4 positive patients tend to have more severe arthritis and extraarticular manifestations, such as vasculitis. Recent studies show that seropositive and seronegative rheumatoid arthritis are associated with DR4. Several studies have suggested a genetic relationship between gold and penicillamine-induced thrombocytopenia and proteinuria and HLA-D region antigens (115). However, the association is not sufficiently strong to contraindicate their use in patients with these antigens. In juvenile rheumatoid arthritis, patients with seropositive polyarticular disease have a high frequency of HLA-DR4 positivity.

Steere et al provided evidence that HLA-DR4 and HLA-DR2 are important in determining susceptibility to arthritis in Lyme disease (116). A significant association between HLA-DR4 and polymyalgia rheumatica has been found in some studies, but not in others (117).

HLA-DR2 and DR3 have been associated with anti-Ro and anti-La in patients with systemic lupus erythematosus and Sjögren's syndrome. However,

C4 null genes have proved to be the most strongly associated genetic factor in systemic lupus erythematosus.

Despite their great theoretical interest, HLA antigens have so far proved of little value to the clinician.

TESTS FOR LYME DISEASE

Lyme disease is caused by the spirochete *Borrelia burgdorferi,* which is transmitted by the deer tick, *Ixodes dammini,* or other Ixodes species. There are two labortory approaches to confirm a diagnosis of Lyme borreliosis. The first is direct demonstration of spirochetes in body fluids or tissue. This is seldom successful, except in the skin. The second, currently the only practical method, is serological determination of antibodies to *B. burgdorferi.* This can be done by indirect immunofluorescent antibody assay, ELISA, and Western blot analysis. All these assays detect IgG and IgM antibodies. Although Western blot analysis is currently the "gold standard" and is the only assay that can define antigen specificity with a high degree of clinical sensitivity and specificity, it is essentially a research tool and not available for routine diagnosis.

Specific IgM antibody titres against *B. burgdorferi,* as determined by immunofluorescence and ELISA methods, reach a peak between 3–6 weeks of illness. In contrast, specific IgG antibody titers rise more slowly and are at their highest some months after the onset of illness (118). It is common for immunofluorescence and ELISA methods to give only weakly positive results. Therefore, diagnosis must rely on clinical findings (119). Only rarely does a patient with Lyme disease have consistently negative tests by these methods. When neuroborreliosis is suspected, cerebrospinal fluid can be tested. Levels of antibody in the serum should be compared and adjusted for differences in total IgG, IgM, or albumin content. The ratio of antibody titer in cerebrospinal fluid and serum can then be used to identify intrathecal synthesis. *Borrelia burgdorferi* has epitopes shared with other spirochetes (e.g., *Treponema pallidum*), but the VDRL test is negative in Lyme disease. Antibodies to *B. burgdorferi* may occur in 50% of patients with nonspirochetal subacute bacterial endocarditis (120).

TESTS FOR RHEUMATIC FEVER

Although still common in poorer parts of the world, acute rheumatic fever has become rare in developed countries (121,122). However, several recent outbreaks have occurred in the United States (123). Diagnosis of acute rheumatic fever is based on clinical features and evidence of a recent streptococcal infection (122). The strain of streptococcus is Group A beta-hemolytic and the site of infection is in the throat. Conventional throat swab cultures are positive in only 25% of patients. Although rapid detection tests for Group A beta-hemolytic streptococci are highly specific, their variable sensitivity makes them unreliable (121,124,125). Confirmation of antecedent infection relies mainly on serum antibody testing.

Streptolysin O is an extracellular toxin produced by Group A beta hemolytic streptococci, which antigenically stimulates production of antistreptolysin O (ASO). Precise quantitation of ASO in IU/mL can now be achieved by rate nephelometry, enabling monitoring of small changes in serum levels. Specificity for antistreptolysin O is ensured by use of recombinant DNA manufacturing, which produces pure streptolysin O antigen, thus eliminating cross reactive antigens, lot-to-lot variation, and exposure to the hemolytic activity of streptolysin O. Generally accepted normal ranges for ASO by rate nephelometry are <200 IU/mL for adults and <100 IU mL for children. Eighty per cent of patients with recent pharyngitis have positive ASO tests. Tests for antibodies to other streptococcal extracellular antigens, such as antihyaluronidase, antistreptokinase, antidiphosphopyridine dinucleotidase, and anti-DNAse-B, may be of value in diagnosis when the ASO test is negative. A commercially available latex agglutination test to detect these antibodies simultaneously is not recommended (121,124,125).

A positive ASO test does not confirm the diagnosis of acute rheumatic fever, but merely that the patient has or has had infection with Group A beta hemolytic streptococci. When chorea is the only presenting symptom, the ASO test may be normal due to the long time elapsed between the infection to the start of symptoms. The original Jones criteria for the diagnosis of rheumatic fever have been revised (126).

REFERENCES

1. Eisenberg JM, Schumacher HR, Davidson PK, et al. Usefulness of synovial fluid analysis in the evaluation of joint effusions. Arch Intern Med 1984;144:715–719.
2. Freemont AJ, Denton J, Chuck A, et al. Diagnostic value of synovial fluid microscopy: a reassessment and rationalization. Ann Rheum Dis 1991;50:101–107.
3. James MJ, Cleland LG, Rofe AM, et al. Intra-articular pressure and the relationship between synovial perfusion and metabolic demand. J Rheumatol 1990;17:521–527.
4. Gatter RA, Schumacher HR. A practical handbook of synovial fluid analysis. 2nd ed. Philadelphia: Lea and Febiger, 1991.
5. Doherty M, Hazelman BL, Hutton CW, et al. Rheumatology examination and injection techniques. London: WB Saunders Co. Ltd., 1992.

6. Gatter RA. Arthrocentesis technique and intrasynovial therapy. In: McCarthy DJ, Koopman WJ, eds. Arthritis and allied conditions. 12th ed. Malvern: Lea and Febiger, 1993;1: 711–720.

7. McCarty DJ. Synovial fluid. In: McCarthy DJ, Koopman WJ, eds. Arthritis and allied conditions: a textbook of rheumatology. 12th ed. Philadelphia: Lea and Febiger, 1993;1:63–84.

8. Hasselbacher P. Arthrocentesis, synovial fluid analysis and synovial biopsy. In: Schumacher HR Jr, Klippel JH, Koopman WJ, eds. Primer on the rheumatic diseases. 10th ed. Atlanta: The Arthritis Foundation, 1993;67–72.

9. Gibson T, Norris W. Skin fragments removed by injection needles. Lancet 1958;2:983–985.

10. Goldman M, Blackman MA. Blood product-associated bacterial sepsis. Transfusion Med Rev 1991;5:73–83.

11. Blackman MA, Ali AM. Bacteria in the blood supply: an overlooked issue in transfusion medicine. In: Nance SJ, ed. Blood safety: current challenges. Bethesda, Maryland: American Association of Blood Banks, 1992;213–228.

12. Field EJ, Harrison RJ. Anatomical terms: their origin and derivation. 2nd ed. Cambridge: Heffer, 1957.

13. Ropes MW, Bauer W. Synovial fluid changes in joint disease. Cambridge: Harvard University Press, 1953.

14. Poulter LW. Immunocytology of synovial fluid cells may be of diagnostic and prognostic value in arthritis. Ann Rheum Dis 1986;45:584–590.

15. Krey RP, Bailen DA. Synovial fluid leukocytosis: a study of extremes. Am J Med 1979;67: 436–442.

16. Hasselbacher P. Variation in synovial fluid analysis by hospital laboratories. Arthritis Rheum 1987;30:637–642.

17. McCutchan MJ, Fisher RC. Synovial leukocytosis in infectious arthritis. Clin Orthop 1990;257:226–230.

18. Klofkorn RW, Lehman TJ. Eosinophilic synovial effusions complicating chronic urticaria and angioedema. Arthritis Rheum 1982;25:708–709.

19. Brown JP, Rola-Pleszczynski M, Menard HA. Eosinophilic synovitis: clinical observations on a newly recognized subset of patients with dermatographism. Arthritis Rheum 1986; 29:1147–1151.

20. Kay J, Eichenfeld AH, Athneya BH, et al. Synovial fluid eosinophilia in Lyme disease. Arthritis Rheum 1988;31:1384–1389.

21. Martin-Santos JM, Mulero J, Andreu JL, et al. Arthritis in idiopathic eosinophilic syndrome. Arthritis Rheum 1988;31:120–125.

22. Gatter RA, Richmond JD. Predominance of synovial fluid lymphocytes in early rheumatoid arthritis. J Rheumatol 1975;2:340–345.

23. Garrido G. A review of peripheral tuberculous arthritis. Semin Arthritis Rheum 1988;18: 142–149.

24. Baker DG, Schumacher HR. Acute monarthritis. N Engl J Med 1993;329:1013–1020.

25. Davison PT, Horowitz J. Skeletal tuberculosis: a review with patient presentations and discussions. Am J Med 1970;48:77–84.

26. Boulware DW, Lopez M, Gum OB. Tuberculous podagra. J Rheumatol 1985;12:1022–1024.

27. Hoffman GS, Myers RL, Stark FR, et al. Septic arthritis associated with mycobacterium avium: a case report and literature review. J Rheumatol 1978;5:199–209.

28. Katzenstein D. Isolated candida arthritis: report of a case and definition of a distinct clinical syndrome. Arthritis Rheum 1985;28:1421–1424.

29. Borenstein DG, Gibbs CA, Jacobs RP. Gas liquid chromatographic analysis of synovial fluid: succinic acid and lactic acid as markers of septic arthritis. Arthritis Rheum 1982;25: 947–953.

30. Cohen AS, Goldenberg D. Synovial fluid. In: Cohen AS, ed. Laboratory diagnostic procedures in the rheumatic diseases. 3rd ed. Philadelphia: Grune and Stratton, 1985;1–54.

31. Shmerling RH. Synovial fluid analysis: a critical reappraisal. Rheum Dis Clin N Am 1994;20:503–512.
32. Pascual E, Tovar J, Ruiz MT. The ordinary light microscope: an appropriate tool for provisional detection and identification of crystals in synovial fluid. Ann Rheum Dis 1989;48: 983–985.
33. Bjelle A. Crystals in joints. Baillière's Clin Rheumatol 1988;2:103–129.
34. McCarty DJ. Crystal identification in human synovial fluids: methods and interpretation. Rheum Dis Clin N Am 1988;14:253–267.
35. Pascual E. Persistence of monosodium urate crystals and low-grade inflammation in the synovial fluid of patients with untreated gout. Arthritis Rheum 1991;34:141–145.
36. Fiechtner JJ, Simkin PA. Urate spherules in gouty synovia. J Am Med Asso 1981;245:1523–1536.
37. Schumacher HR, Reginato AJ. Atlas of synovial fluid analysis and crystal identification. Philadelphia: Lea and Febiger, 1991.
38. Ryan LM. Calcium pyrophosphate crystal deposition disease; pseudogout; articular chondrocalcinosis. In: McCarthy DJ, Koopman WJ, eds. Arthritis and allied conditions. 12th ed. Philadelphia: Lea and Febiger, 1993;2:1835–1872.
39. Paul H, Reginato AJ, Schumacher HR. Alizarin red S staining as a screening test to detect calcium compounds in synovial fluids. Arthritis Rheum 1983;26:191–200.
40. Reginato AJ, Schumacher HR, Allan DA, et al. Acute monarthritis associated with lipid liquid crystals. Ann Rheum Dis 1985;44:537–543.
41. Hoffman GS, Schumacher HR, Paul H, et al. Calcium oxalate microcrystalline-associated arthritis in end-stage renal disease. Ann Intern Med 1982;97:36–42.
42. Wallis WJ, Simkin PA. Antirheumatic drug concentrations in human synovial fluid and synovial tissue: observations on extravascular pharmacokinetics. Clin Pharmacokinet 1983;8:496–522.
43. Steinbrocker O, Neustadt DH. Aspiration and infection therapy in arthritis and musculoskeletal disorders. Hagerstown, MD: Harper and Row, 1972.
44. Schumacher HR, Kulka JP. Needle biopsy of the synovial membrane: experience with the Parker-Pearson technique. N Engl J Med 1972;8:416–419.
45. Johnson LL. Diagnostic and surgical arthroscopy: the knee and other joints. St Louis: CV Mosby Co., 1981.
46. Altman RD, Kates J. Arthroscopy of the knee. Semin Arthritis Rheum 1983;13:188–199.
47. Huff JP, Sequeira W, Harris CA, et al. Survey of physicians doing office-based arthroscopy. Arthritis Rheum 1992;35(suppl):S292.
48. Palmer DG, Schumacher HR. Synovitis with nonspecific histologic changes in synovium in chronic sarcoidosis. Ann Rheum Dis 1984;43:778–782.
49. Iguchi T, Matsubara T, Kawai K, et al. Clinical and histologic observations and monarthritis: anticipation of its progression to rheumatoid arthritis. Clin Orthop 1990;250:241–249.
50. Sissons HA, Murray RO, Kemp HBS. Orthopaedic diagnosis: clinical, radiological and pathological considerations. Berlin: Springer-Verlag, 1984.
51. Rosa PA, Schwan TG. A specific and sensitive assay for the Lyme disease spirochete Borrelia burgdorferi using the polymerase chain reaction. J Infect Dis 1989;160:1018–1029.
52. Anderson JR. Atlas of skeletal muscle pathology. Lancaster: MTP Press Ltd., 1985.
53. Dalakas MC. Polymyositis, dermatomyositis, and inclusion body myositis. N Engl J Med 1991;325:1487–1498.
54. Morton BD III, Statland BE. Serum enzyme alterations in polymyositis. Am J Clin Pathol 1980;73:556–557.
55. Bernstein RM, Morgan SH, Chapman J, et al. Anti-Jo-1 antibody: a marker for myositis with interstitial lung disease. Br Med J 1984;289:151–152.
56. Fraser DD, Frank JA, Dalakas M, et al. Magnetic resonance imaging in the idiopathic inflammatory myopathies. J Rheumatol 1991;18:1693–1700.

57. Wortmann RL. Myositis or myopathy. J Rheumatol 1989;16:1525–1527.
58. Valen PA, Nakayama DA, Veum J, et al. Myoadenylate deaminase deficiency and forearm ischemic exercise testing. Arthritis Rheum 1987;30:661–668.
59. Engel WK, Eyerman EL, Williams HE. Late-onset type of skeletal muscle phosphorylase deficiency: a new familial variety with completely and partially affected subjects. N Engl J Med 1963;268:135–137.
60. Otake S, Banno T, Onba S, et al. Muscular sarcoidosis: findings at MR imaging. Radiology 1990;176:145–148.
61. Fernandes L, Swinson DR, Hamilton EBD. Dermatomyositis complicating penicillamine treatment. Ann Rheum Dis 1977;36:94–95.
62. Blakeman JM. The punch skin biopsy. Canad Fam Phys 1983;29:971–974.
63. Lever WF, Schaumburg-Lever G. Histopathology of the skin. 6th ed. Philadelphia: JB Lippincott Co., 1983.
64. Mackie RM. Milne's dermatopathology. Revised 2nd ed. London: Edward Arnold, 1984.
65. McCluskey RT. The value of renal biopsy in lupus nephritis. Arthritis Rheum 1982;25:868–875.
66. Austin HA, Muenz LR, Joyce KM, et al. Prognostic factors in lupus nephritis: contribution of renal histologic data. Am J Med 1983;75:382–391.
67. Cupps TR, Fauci AS. The vasculitides. In: Smith LH, ed. Major problems in internal medicine. Philadelphia: WB Saunders Co., 1981.
68. Churg A, Churg J, eds. Systemic vasculitides. New York: Igaku-Shoin. Medical Publishers, Inc., 1991.
69. Chisholm DM, Mason DK. Labial salivary gland biopsy in Sjögren's disease. J Clin Path 1968;21:656–660.
70. Greenspan JS, Daniels TE, Talal N, et al. The histopathology of Sjögren's syndrome in labial salivary gland biopsies. Oral Surg 1974;37:217–229.
71. Tarplay TM Jr, Anderson LG, White CL. Minor salivary gland involvement in Sjögren's syndrome. Oral Surg 1974;37:64–74.
72. Fox RI, Kang HI. Pathogenesis of Sjögren's syndrome. Rheum Dis Clin North Am 1992;18:517–538.
73. Gaensler EA, Carrington CB. Open biopsy for diffuse infiltrative lung disease: clinical, roentgenographic and histological correlations in 502 patients. Ann Thoracic Surg 1980;30:411–425.
74. Gaeve AH, Saul VA, Bechara FA. Role of different methods of lung biopsy in the diagnosis of lung lesions. Am J Surg 1980;140:742–746.
75. Sterret G, Whitaker D, Glancy J. Fine-needle aspiration of lung, mediastinum and chest wall. Path Anat 1982;17:197–225.
76. Baer AN, Dessypris EN, Krantz SB. The pathogenesis of anemia in rheumatoid arthritis: a clinical and laboratory analysis. Semin Arthr Rheum 1990;19:209–223.
77. Vreugdenhill G, Baltus CA, van Eijk HG, et al. Anemia of chronic disease: diagnostic significance of erythrocyte and serological parameters of iron deficient rheumatoid arthritis patients. Brit J Rheumatol 1990;29:105–110.
78. Bjarnason I. Nonsteroidal antiinflammatory drug-induced small intestinal inflammation in man. In: Pounder RE, ed. Recent advances in gastroenterology. No 7. Edinburgh: Churchill Livingstone, 1988;23–46.
79. Wyngaarden JB, Kelley WN. Gout and hyperuricemia. New York: Grune and Stratton, 1976.
80. Hall AP, Barry PE, Dawber TR, et al. Epidemiology of gout and hyperuricemia: a long-term population study. Am J Med 1967;42:27–37.
81. Price CP, James DR. Analytical reviews in clinical biochemistry: the measurement of urate. Am Clin Biochem 1988;25:484–498.
82. Rieselbach RE, Steele TH. Influence of the kidney upon urate homeostasis in health and disease. Am J Med 1974;56:665–675.

83. Fessel JW. Renal outcomes of gout and hyperuricemia. Am J Med 1979;67:74–82.
84. Madhor R, Crilly A, Watson J, et al. Serum interleukin-6 levels in rheumatoid arthritis: correlations with clinical and laboratory indices of disease activity. Ann Rheum Dis 1993;52:232–234.
85. Pepys MB, Lanham JG, De Beer FC. C-reactive protein in SLE. Clin Rheum Dis 1982;8:91–103.
86. Caswell M, Pike LA, Bull BS, et al. Effect of patient age on tests of the acute-phase response. Arch Path Lab Med 1993;117:906–910.
87. Khawar K, Al-Jarallah K, Buchanan WW. Cyclosporin A in rheumatoid arthritis: a critical review. Inflammopharmacol 1993;2:141–157.
88. Dawes PT, Fowler PD, Clarke S, et al. Rheumatoid arthritis: treatment which controls the C-reactive protein and erythrocyte sedimentation rate reduces radiological progression. Br J Rheumatol 1986;25:44–49.
89. Dixon JS. Relationship between plasma viscosity or ESR and the Ritchie articular index. Br J Rheumatol 1984;23:233–235.
90. Bilezikian JP. Hypercalcemic states, their differential diagnosis and acute management. In: Coe FD, Favus MJ, eds. Disorders of bone and mineral metabolism. New York: Raven Press, 1992;492.
91. Eastell R, Heath H III. The hypocalcemic states, their differential diagnosis and acute management. In: Coe FD, Favus MJ, eds. Disorders of bone and mineral metabolism. New York: Raven Press, 1992;572.
92. Cockroft DW, Gault MH. Prediction of creatinine clearance from serum creatinine. Nephron 1976;16:31–41.
93. Flynn FW. Assessment of renal function: selected developments. Clin Biochem 1990;23: 49–54.
94. O'Brien WM, Bagby GF. Rare adverse reactions to nonsteroidal antiinflammatory drugs. J Rheumatol 1985;12:13–20, 347–353, 562–567, 785–790.
95. Teodorescu M, Froelich CJ. Detection and measurement of rheumatoid factor. In: Teodorescu M, Froelich CJ, eds. Advanced immunoassays in rheumatology. Boca Raton: CRC Press, 1994;51–60.
96. Koopman WJ, Schrohenioher RE. Rheumatoid factor: mechanisms of production and biological significance. In: Teodorescu M, Froelich CJ, eds. Advanced immunoassays in rheumatology. Boca Raton: CRC Press, 1994;29–50.
97. Harris J, Vaughan JH. Transfusion studies in rheumatoid arthritis. Arthritis Rheum 1961;4:47–55.
98. Vaughan JH. 1992 Joseph J. Bunim Lecture: pathogenetic concepts and origins of rheumatoid factor in rheumatoid arthritis. Arthr Rheum 1993;36:1–6.
99. Torrigiani G, Roitt IM. Antiglobulin factors in sera from patients with rheumatoid arthritis and normal subjects: quantitative estimation in different immunoglobulin classes. Ann Rheum Dis 1967;26:334–340.
100. Roitt IM, Sumar N. IgG and rheumatoid factor at a glance. Clin Exper Rheumatol 1990;8:89–91.
101. Aho K, Heliovaara M, Maatela J, et al. Rheumatoid factors antedating clinical rheumatoid arthritis. J Rheumatol 1991;18:1282–1284.
102. Froelich CJ, Teodorescu M. Clinical significance and interpretation of antinuclear antibody tests. In: Teodorescu M, Froelich CJ, eds. Advanced immunoassays in rheumatology. Boca Raton: CRC Press, 1994;123–157.
103. Froelich CJ, Lahita RG. Nuclear and cytoplasmic antigens and induction of autoantibodies. In: Teodorescu M, Froelich CJ, eds. Advanced immunoassays in rheumatology. Boca Raton: CRC Press, 1994;78–101.
104. Teodorescu M, Froelich CJ. Detection and measurement of antinuclear and anticytoplasmic antibodies. In: Teodorescu M, Froelich CJ, eds. Advanced immunoassays in rheumatology. Boca Raton: CRC Press, 1994;103–121.

105. Harris EN, Exner T, Hughes GRV, et al. Phospholipid-binding antibodies. Boca Raton: CRC Press, 1991.
106. Barna LK, Triplett DA. A report on the first international workshop for lupus anticoagulant identification. Clin Exper Rheumatol 1991;9:557–567.
107. Asherson RA, Cervera R. The antiphospholipid syndrome: a syndrome in evolution. Ann Rheum Dis 1992;51:147–150.
108. Hess EV. Drug-related lupus: the same or different? In: Lahita RG, ed. Systemic lupus erythematosus. New York: Wiley Medical, 1987;869–880.
109. Morgan PB. Complement clinical aspects and relevance to disease. London: Academic Press, 1990.
110. Roitt I, Brostoff J, Male D. Immunology. London: Gower Medical Publishing Ltd., 1985.
111. Brostoff J, Scadding GK, Male D, et al. Clinical immunology. London: Gower Medical Publishing, 1991.
112. Wallace DJ, Hahn BH, eds. Dubois' lupus erythematosus. 4th ed. Philadelphia: Lea and Febiger, 1993.
113. Al-Jarallah K, Singal D, Buchanan WW. Human leukocyte antigens (HLA) and rheumatic disease: HLA class I antigen associated diseases. Inflammopharmacology 1993;2:37–45.
114. Singal DP, Buchanan WW. Human leukocyte antigens (HLA) and rheumatic disease: HLA class I antigen associated diseases. Inflammopharmacology 1993;2:47–62.
115. Singal DP, Green D, Reid B, et al. HLA-D region genes and rheumatoid arthritis (RA): importance of DR and DQ genes in conferring susceptibility to RA. Ann Rheum Dis 1992;51:23–28.
116. Steere AC, Dwyer E, Winchester R. Association of chronic Lyme arthritis with HLA-DR4 and HLA-DR2 alleles. N Engl J Med 1990;323:219–223.
117. Al-Jarallah KF, Buchanan WW, Sastry A, et al. Immunogenetics of polymyalgia rheumatica. Clin Exper Rheumatol 1993;11:529–531.
118. Schoen RT. Identification of Lyme disease. Rheum Dis Clinics N Am 1994;20:361–369.
119. Shrestha M, Grodzicki RI, Steere AC. Diagnosing early Lyme disease. Am J Med 1985;78:235–240.
120. Kaell AT, Redecha PR, Elkon KB. Occurrence of antibodies to Borrelia burgdorferi in patients with nonspirochetal subacute bacterial endocarditis. Ann Intern Med 1993;119: 1079–1083.
121. Bisno AL. Group A streptococcal infections and acute rheumatic fever. N Engl J Med 1991;325:783–793.
122. Taranta A, Markowitz M. Rheumatic Fever. 2nd ed. Dordrecht: Kluwer Academic, 1989.
123. Veasey LG, Wiedmeir SE, Orsmond GS, et al. Resurgence of acute rheumatic fever in the intermountain area of the USA. N Engl J Med 1987;316:421–427.
124. Prevention of rheumatic fever: a statement for health professionals by the Committee on Rheumatic Fever, Endocarditis, and Kawasaki Diseases of the Council on Cardiovascular Disease in the Young. The American Heart Association Circulation 1988;78:1082–1086.
125. Rheumatic fever and rheumatic heart disease: report of a World Health Study Group. WHO Tech Rep Ser 1988;764:1–58.
126. Stollerman GH, Markowitz M, Taranta A, et al. Jones criteria (revised) for guidance in the diagnosis of rheumatic fever. Circulation 1965;32:664–668.

Appendix

Carolus Linnaeus (1701–1778) was a botanist and a doctor. After organizing the taxonomy of the plant world, he applied his skills to differentiating human disease based on the patient's symptoms. Today, clinicians continue to base diagnosis on symptoms and signs. In addition, they make use of a wide range of laboratory data, including bacteriology, biochemistry, immunology, virology, histopathology, and radiology. Due to rapid developments in laboratory medicine, nomenclature and classification of disease continue to change. This is particularly true of rheumatic disease.

Although diagnosis is the cornerstone of medical practice, its definition and ascertainment is less clear. Many strategies have been used in making a diagnosis. Defining probabilities, although of interest to statisticians, is probably rare in clinical practice. Although algorithms are useful teaching aids and provide therapeutic guidelines, they are not popular with clinicians. We believe that diagnosis is essentially a gestalt process of pattern recognition. Modern computers do not have strong pattern recognition capabilities.

RATING INDICES: SENSITIVITY, SPECIFICITY, AND PREDICTIVE VALUES

Clinicians should be familiar with the terms used in predicting the value of tests. The sensitivity of a test is defined as the percentage of persons with the disease who are positive. The formula derived from the contingency table (Appendix Table 1) is:

Table 1. Predictive Value of Tests

Results of Diagnostic Test	True Status of Disease		Totals
	Positive	Negative	
Positive	True Positive (A)	False Positive (B)	A + B
Negative	False Negative (C)	True Negative (D)	C + D
Total	(A + C)	(B + D)	(A + B + C + D)

$$\text{Sensitivity} = \frac{\text{true positive (A)}}{\text{true positive (A)} + \text{false negative (C)}}$$

Specificity, on the other hand, refers to the percentage of nondiseased persons who are negative.

$$\text{Specificity} = \frac{\text{true negative (D)}}{\text{false positive (B)} + \text{true negative (D)}}$$

The formulae for other rating indices are as follows:

$$\text{Positive predictive value} = \frac{\text{true positive (A)}}{\text{true positive (A)} + \text{false positive (B)}}$$

$$\text{Negative predictive value} = \frac{\text{true negative (D)}}{\text{true negative (D)} + \text{false negative (C)}}$$

$$\text{Likelihood ratio} = \frac{\text{Sensitivity}}{1 - \text{specificity}}$$

$$\text{Relative risk} = \frac{\text{true positive (A)} \times \text{true negative (D)}}{\text{false positive (B)} \times \text{false negative (C)}}$$

$$\text{Pretest disease probability} = \frac{\text{true positive (A)} \times \text{false negative (B)}}{\text{true positive (A)} + \text{false negative (C)} + \text{false positive (B)} + \text{true negative (D)}}$$

$$\text{Posttest disease probability with positive test} = \frac{\text{true positive (A)}}{\text{true positive (A)} + \text{false positive (C)}}$$

$$\text{Posttest disease probability with negative test} = \frac{\text{false negative (C)}}{\text{false negative (C)} + \text{false positive (D)}}$$

For the interested reader, Lequesne and Wilhelm produced a useful compendium and glossary of statistical methodology (1).

An easily read guide to SI unit conversion has been published by Laposata (2). The conversion factors from SI to standard units, and vice versa, for laboratory tests commonly used by clinical rheumatologists are summarized in Appendix Table 2. Normal ranges are shown because they differ in different laboratories.

References to diagnostic criteria for different rheumatic diseases are summarized in Appendix Table 3.

CLINICAL THERAPEUTIC TRIALS

The fundamental issues and various outcome measures used in clinical therapeutic trials in various forms of arthritis and rheumatism have been well described (3,4).

Current agency guidelines for efficacy assessments in rheumatoid arthritis include those of the U.S. Food and Drug Administration (FDA) and of the European League Against Rheumatism (EULAR) (5,6). The guidelines of the FDA include:

1. Number of painful or tender joints
2. Number of swollen joints
3. Duration of morning stiffness
4. Grip strength
5. 50-foot walk time
6. Erythrocyte sedimentation rate (ESR)
7. Principal observer's opinion of patient's condition
8. Patient's opinion of the condition

The EULAR guidelines include:

1. Measurement of pain on movement or during weight bearing using a visual analogue (VA) or 4-point scale
2. Rest pain or night pain (number of times awakened by pain during the night)
3. Duration of morning stiffness

Table 2. Traditional and SI Units and Conversion Factors

Laboratory Test	Units Traditional	SI	Conversion Factor
Biochemical			
Serum acid phosphatase (prostatic)	U/L	U/L	1
Serum albumin	g/dL	q/L	10
Serum alkaline phosphatase	Units/L	U/L	1
Serum ALT*	Units/L	U/L	1
Serum amylase	Units/L	U/L	1
Serum AST†	Units/L	U/L	1
Serum bilirubin	mg/dL	μmol/L	17.1
Blood urea nitrogen	mg/dL	mmol/L	0.357
Serum calcium	mg/dL	mmol/L	0.250
Serum ionized calcium	mEq/L	mmol/L	0.500
Urinary calcium	mg/24 hr	mmol/d	0.02495
Serum creatine kinase†	Units/L	U/L	1
Serum creatinine	mg/dL	μmol/L	88.4
Serum immunoglobulins (G,A,M)	mg/dL	q/L	0.01
Serum phosphorus (inorganic phosphate)	mg/dL	mmol/L	0.3229
Serum total protein	g/dL	g/L	10
Serum urate	mg/dL	μmol/L	59.48
Urinary urate	g/24 hr	mmol/d	5.948
Hematologic			
Hemoglobin	q/dL	q/L	10
Mean corpuscular hemoglobin	pg	pg	1
Mean corpuscular hemoglobin concentration	q/dL	g/L	10
Mean corpuscular volume	μm^3	fL	1
Red cell count	$10^6/mm^3$	$10^{12}/L$	1
White cell count	$10^3/mm^3$	$10^9/L$	1
Differential white cell count	(cells/mm^3)	$10^6/L$	1
Platelet count	$10^3/mm^3$	$10^9/L$	1
Serum ferritin	ng/ml	uq/L	1
Serum iron	μg/dL	μmol/L	0.1791
Serum iron binding capacity	μq/dL	μmol/L	0.1791
Erythrocyte sedimentation rate	mm/hr	mm/hr	1
Immumologic			
Serum complement (C3 and C4)	mq/dL	q/L	0.01
Pharmacologic			
Serum acetaminophen§	mg/dL	μmol/L	66.16
Serum salicylate‖	mg/dL	mmol/L	0.07240

Conversion factor: multiply from traditional to SI unit, and divide from SI to traditional units.
* ALT (alanine aminotransferase, SGPT)
† AST (aspartate aminotransferase, SGOT)
† Creatine kinase (creatine phosphokinase)
§ Acetaminophen: therapeutic serum concentrations 0.2–0.6 mg/dL or 13–40 μmol/L and toxic serum concentration 75.0 mg/dL or >330 μmol/L
‖Salicylate: salicylate therapeutic serum concentrations 20–30 mg/dL (1.45–2.17 mmol/L)

Table 3. Diagnostic Criteria for Different Rheumatic Diseases

Disease	Author and Year	Reference #
Osteoarthritis		
Clinical criteria	Altman, et al, 1987	8
Knee	Altman, et al, 1986	9
Hand	Altman, et al, 1990	10
Hip	Altman, et al, 1990	11
Radiograph assessment	Altman, et al, 1987	12
Rheumatoid arthritis	Ropes, et al, 1956	13
	Ropes, et al, 1958	14
	Arnett, et al, 1988	15
Juvenile rheumatoid arthritis	Cassidy, et al, 1989	16
Sjögren's syndrome	Manthorpe, et al, 1986	17
	Skopouli, et al, 1986	18
	Homma, et al, 1986	19
	Fox, et al, 1986	20
	Vitali, et al, 1993	21
	For discussion of criteria	22
Ankylosing Spondylitis		
Rome criteria	Kelgren, 1962	23
New York criteria	Vennett and Wood, 1968	24
	Khan and van der Linden, 1990	25
	Zeidler, et al, 1992	26
Psoriatic arthritis	Moll and Wright, 1973	27
Reiter's disease	Wilkens, et al, 1981	28
Behçet's disease	International Study Group, 1990	29
Systemic lupus erythematosus	Tan, et al, 1982	30
Progressive systemic sclerosis	Subcommittee of the American Rheumatism Association, 1980	31
Dermatomyositis and polymyositis	Bohan and Peter, 1975	32
Raynaud's phenomenon	LeRoy and Medsger, 1992	33
Vasculitis		
Polyarteritis Nodosa	Lightfoot, et al, 1990	34
Churg-Strauss syndrome	Masi, et al, 1990	35
Wegener's granulomatosis	Leavitt, et al, 1990	36
Hypersensitivity vasculitis	Calabrese, et al, 1990	37
Henoch-Schönlein purpura	Mills, et al, 1990	38
Giant cell arteritis	Hunder, et al, 1990	39
Takayasu's arteritis	Arend, et al, 1990	40
General discussion on classification	Hunder, et al, 1990	41
	Bloch, et al, 1990	42
	Fries, et al, 1990	43
	Mandell and Hoffman, 1994	44
Histopathologic classification	Lie, et al, 1990	45
Gout	Wallace, et al, 1977	46
Calcium pyrophosphate dehydrate crystal deposition disease	Ryan and McCarty, 1993	47

Table 3.—continued

Disease	Author and Year	Reference #
Rheumatic fever	Stollerman, et al, 1990	48
Fibromyalgia	Wolfe, et al, 1990	49
Hypertrophic osteoarthropathy	First International Workshop, 1992	50

4. Grip strength
5. Measurement of proximal interphalangeal joint circumference
6. Articular index (e.g., Ritchie)
7. Functional index (e.g., Lee)
8. Overall evaluation by patient in comparison to pretrial state
9. Overall evaluation by physician

An excellent review of health status measurement can be found in the third volume of the *Dictionary of the Rheumatic Diseases* (7).

CLASSIFICATION OF ANTIRHEUMATIC DRUGS

A panel of American, Australian, and English rheumatologists recently proposed a new classification of antirheumatic drugs (51). Because all antirheumatic drugs are effective in relieving symptoms, it is proposed that they be classified as Symptom-Modifying Antirheumatic Drugs (SM-ARDs). These include the nonsteroidal antiinflammatory drugs (NSAIDs) [category I], corticosteroids (category II), and the slower acting medications, such as antimalarials, gold, D-penicillamine, antimetabolites, and cytotoxic agents (category III). They should only be referred to as disease-controlling antirheumatic agents (DC-ARTs) if one of the drugs or combination of such drugs improves and sustains function in association with reduction in joint inflammation and prevents or significantly decreases the rate of radiologic joint damage for at least 1 year. It is suggested that the time period for which these criteria are satisfied should be recorded (e.g., 3-year DC-ARTs). To our knowledge, no antirheumatic drug singly or in combination warrants the classification of DC-ARTs.

REFERENCES

1. Lequesne M, Wilhelm F. Methodology for the clinician: compendium and glossary. Basel: EULAR Publishers, 1989.

2. Laposata M. SI unit conversation guide. Boston: NEJM Books, 1992.
3. Bellamy N, Buchanan WW. Clinical evaluation in rheumatic diseases. In: McCarty DJ, Koopman WJ, eds. Arthritis and allied conditions. 12th ed. Philadelphia: Lea and Febiger, 1993;151–178.
4. Bellamy N. Musculoskeletal clinical methodology. Dordrecht: Kluwer Academic Publishers, 1993.
5. Guidelines for the clinical evaluation of antiinflammatory and antirheumatic drugs (adults and children). Washington, District of Columbia: US Department of Health and Human Services. Public Health Services. Food and Drug Administration, 1988;2–11, 29–35.
6. Guidelines for the clinical investigation of drugs used in rheumatic diseases: European drug guidelines, Series 5. Copenhagen: WHO, Regional Office for Europe. European League Against Rheumatism, March 1985;7–11.
7. Dictionary of the Rheumatic Diseases, Vol III. Health Status Measurement. Prepared by the Glossary Committee of the American College of Rheumatology. Bayport, New York: Contact Associates International Ltd., 1988.
8. Altman RD, Bloch DA, Bole GG Jr, et al. Development of clinical criteria for osteoarthritis. J Rheumatol 1987;14(Suppl):3–6.
9. Altman RD, Asch A, Bloch DA. Development of criteria for the classification and reporting of osteoarthritis: classification of osteoarthritis of the knee. Arthritis Rheum 1986;29:1039–1049.
10. Altman RD, Alarcon G, Appelrouth D, et al. The American college of rheumatology criteria for the classification and reporting of osteoarthritis of the hand. Arthritis Rheum 1990;33:1601–1610.
11. Altman RD, Alarcon G, Appelrouth D, et al. The American college of rheumatology criteria for the classification and reporting of osteoarthritis of the hip. Arthritis Rheum 1991;34:505–514.
12. Altman RD, Fries JF, Bloch DA. Radiographic assessment of progression in osteoarthritis. Arthritis Rheum 1987;30:1214–1225.
13. Ropes MW, Bennett GA, Cobb S, et al. Proposed diagnostic criteria for rheumatoid arthritis. Bull Rheum Dis 1956;7:121–124.
14. Ropes MW, Bennett GA, Cobb S, et al. 1958 revision of diagnostic criteria for rheumatoid arthritis. Bull Rheum Dis 1958;9:175–176.
15. Arnett FC, Edworthy SM, Bloch DA, et al. The American rheumatism 1987: revised criteria for the classification of rheumatoid arthritis. Arthritis Rheum 1988;31:315–324.
16. Cassidy JT, Levinson JE, Brewer EJ Jr. The development of classification criteria for children with juvenile rheumatoid arthritis. Bull Rheum Dis 1989;38:1–7.
17. Manthorpe R, Oxholm P, Prause JU, et al. The Copenhagen criteria for Sjögren's syndrome. Scand J Rheumatol 1986;61(Suppl):19–21.
18. Skopouli FN, Drusos AA, Papaioannou T, et al. Preliminary diagnostic criteria for Sjögren's syndrome. Scand J Rheumatol 1986;61(Suppl):26–27.
19. Homma M, Tojo T, Akizuki M, et al. Criteria for Sjögren's syndrome in Japan. Scand J Rheumatol 1986;61(Suppl):26–27.
20. Fox RI, Robinson CA, Curd JG, et al. Sjögren's syndrome: proposed criteria for classification. Arthritis Rheum 1986;29:577–585.
21. Vitali C, Bombardieri S, Haralampos M, et al. Preliminary criteria for the classification of Sjögren's syndrome: results of a prospective concerted action supported by the European Community. Arthritis Rheum 1993;36:340–347.
22. Editorial. Diagnosis of Sjögren's syndrome. Lancet 1992;340:150–151.
23. Kellgren JH. Diagnostic criteria for population studies. Bull Rheum Dis 1962;13:291–292.
24. Bennett PH, Wood PHN, eds. Population studies of the rheumatic diseases (New York). Amsterdam: Excerpta Medica Foundation, 1968;148:4–7.
25. Khan MA, van der Linden SM. Ankylosing spondylitis and associated diseases. Rheum Dis Clin N Am 1990;16:551–579.

26. Zeidler H, Mau W, Khan MA. Undifferentiated spondyloarthropathies. Rheum Dis Clin N Am 1992;18:187–202.
27. Moll JMH, Wright V. Psoriatic arthritis. Semin Arthritis Rheum 1973;3:55–78.
28. Wilkens RF, Arnett FC, Bitter T, et al. Reiter's syndrome: evaluation of preliminary criteria for definite disease. Arthritis Rheum 1981;24:844–849.
29. International Study Group for Behçet's Disease. Criteria for diagnosis of Behçet's disease. Lancet 1990;33:1078–1080.
30. Tan EM, Cohen AS, Fries JF, et al. The 1982 revised criteria for the classification of systemic lupus erythematosus. Arthritis Rheum 1982;25:1271–1277.
31. Subcommittee for Scleroderma Criteria of the American Rheumatism Association Diagnostic and Therapeutic Criteria Committee. Preliminary criteria for the classification of systemic sclerosis (scleroderma). Arthritis Rheum 1980;23:581–590.
32. Bohan A, Peter JB. Polymyositis and dermatomyositis. N Engl J Med 1975;292:344–347; 403–407.
33. LeRoy EC, Medsger TA. Raynaud's phenomenon: a proposal for classification. Clin Exper Rheumatol 1992;10:485–488.
34. Lightfoot RW Jr, Beat MA, Bloch DA, et al. The American college of rheumatology criteria for the classification of polyarteritis nodosa. Arthritis Rheum 1990;33:1088–1093.
35. Masi AT, Hunder GG, Lie JT, et al. The American college of rheumatology criteria for the classification of Churg-Strauss Syndrome (allergic granulomatosis and angiitis). Arthritis Rheum 1990;33:1094–1100.
36. Leavitt RY, Fauci AS, Bloch DA, et al. The American college of rheumatology criteria for the classification of Wegener's granulomatosis. Arthritis Rheum 1990;33:1101–1107.
37. Calabrese LH, Michel BA, Broch DA, et al. The American college of rheumatology criteria for the classification of hypersensitivity vasculitis. Arthritis Rheum 1990;33:1108–1113.
38. Mills JA, Michel BA, Bloch DA, et al. The American college of rheumatology criteria for the classification of Henoch-Schönlein purpura. Arthritis Rheum 1990;33:1114–1121.
39. Hunder GG, Bloch DA, Michel BA, et al. The American college of rheumatology criteria for the classification of giant cell arteritis. Arthritis Rheum 1990;33:1122–1128.
40. Arend WP, Michel BA, Bloch DA, et al. The American college of rheumatology criteria for the classification of Takayasu arteritis. Arthritis Rheum 1990;33:1129–1134.
41. Hunder GG, Arend WP, Bloch DA, et al. The American college of rheumatology criteria for the classification of vasculitis: introduction. Arthritis Rheum 1990;33:1065–1067.
42. Bloch DA, Michel BA, Hunder GG, et al. The American college of rheumatology criteria for the classification of vasculitis: patients and methods. Arthritis Rheum 1990;33:1068–1073.
43. Fries JF, Hunder GG, Bloch DA, et al. The American college of rheumatology criteria for the classification of vasculitis: summary. Arthritis Rheum 1990;33:1135–1136.
44. Mandell BF, Hoffman GS. Differentiating the vasculitides. Rheum Dis Clin N Am 1994;20:409–442.
45. Lie JT, Members and Consultants of the American College of Rheumatology Subcommittee on Classification of Vasculitis. Illustrated histopathologic classification of criteria for selected vasculitis syndromes. Arthritis Rheum 1990;33:1074–1087.
46. Wallace Sl, Robinson H, Masi AT, et al. Preliminary criteria for the classification of the acute arthritis of primary gout. Arthritis Rheum 1977;20:895–900.
47. Ryan LM, McCarty DJ. Diagnostic criteria for calcium pyrophosphate dihydrate crystal deposition disease (pseudogout). In: McCarty DJ, Koopman WJ, eds. Arthritis and allied diseases. 12th ed. 1993;2:1835–1855.
48. Stollerman GH, Markowitz M, Taranta A, et al. Jones criteria (revised) for guidance in the diagnosis of rheumatic fever. Circulation 1965;32:664–668.
49. Wolfe F, Smythe HA, Yunus MB, et al. The American College of Rheumatology 1990 Cri

teria for the Classification of Fibromyalgia: report of the multicenter criteria committee. Arthritis Rheum 1990;33:160–172.

50. First International Workshop on Hypertrophic Osteoarthropathy. Clin Exper Rheumatol 1992;7(Suppl):1–71.

51. Edmonds JP, Scott DL, Furst DE, et al. Antirheumatic drugs: a proposed new classification. Editorial. Arthritis Rheum 1993;36:336–339.

Index

References in *italics* denote figures; "t" denotes tables